SHEAR BEHAVIOUR OF ROCK JOINTS

Shear Behaviour
of Rock Joints

BUDDHIMA INDRARATNA & ASADUL HAQUE

Department of Civil, Mining and Environmental Engineering,
University of Wollongong, Australia

A.A.BALKEMA/ROTTERDAM/BROOKFIELD/2000

Published by
A.A. Balkema, P.O. Box 1675, 3000 BR Rotterdam, Netherlands
Fax: +31.10.4135947; E-mail: balkema@balkema.nl; Internet site: http://www.balkema.nl

A.A. Balkema Publishers, Old Post Road, Brookfield, VT 05036-9704, USA
Fax: 802.276.3837; E-mail: info@ashgate.com

ISBN 90 5809 307 7 hardbound edition
ISBN 90 5809 308 5 student paper edition

Contents

PREFACE   IX

1 CHARACTERIZATION OF JOINT SURFACE ROUGHNESS   1
  1.1 Introduction   1
  1.2 Roughness measurement techniques   2
       1.2.1 JRC measurement   2
       1.2.2 Fractal method   4
       1.2.3 Spectral method   9
       1.2.4 Digital coordinate measuring machine   11
       1.2.5 Fourier transform method and its application to measure
              surface rougness   14
  1.3 Physical and mechanical properties of intact and jointed rocks   15
  References   16

2 SHEAR BEHAVIOUR OF CLEAN ROCK JOINTS   17
  2.1 Introduction   17
  2.2 CNS and CNL concepts   18
  2.3 Shear apparatus for laboratory testing   21
       2.3.1 Large scale shear boxes   21
       2.3.2 Loading device   21
       2.3.3 Measurements of displacements   22
  2.4 Laboratory modelling of rock joints   23
       2.4.1 Selection of model material for joint   23
       2.4.2 Preparation of saw-tooth and natural specimens   23
       2.4.3 Sampling and preparation of field joint specimens   24
  2.5 Laboratory testing of rock joints   25
       2.5.1 Shear strength response under Constant Normal Stiffness   26
       2.5.2 Normal stress and dilation behaviour   27
       2.5.3 Effect of normal stiffness on joint shear behaviour   28
       2.5.4 Effect of stiffness on shear displacement corresponding to
              peak shear stress   28
       2.5.5 Effect of shear rate on the strength of joint   28

2.6 Tests on soft rock joints   29
    2.6.1 Effect of shear displacement rate   29
    2.6.2 Effect of boundary condition on shear behaviour   31
    2.6.3 Behaviour of unfilled/clean regular joints under CNS   36
    2.6.4 Shear behaviour of natural (field) joints   37
    2.6.5 Stress-path response of Type I, II and III joints   42
    2.6.6 Strength envelopes for Types I, II and III   44
2.7 Empirical models for the prediction of shear strength of rock joints   45
2.8 Summary of behaviour of unfilled/clean joints   49
    2.8.1 Effect of shear rate on shear behaviour of joints under CNS   49
    2.8.2 Effect of boundary condition on shear behaviour   50
    2.8.3 Shear behaviour of soft unfilled joint under CNS   51
References   51

3 INFILLED ROCK JOINT BEHAVIOUR   54
3.1 Influence of infill on rock joint shear strength   54
3.2 Factors controlling infilled joint shear strength   55
    3.2.1 Effect of joint type on shear behaviour   56
    3.2.2 Infill type and thickness   56
    3.2.3 Effect of drainage condition   63
    3.2.4 Infill boundary condition   63
    3.2.5 Infill-rock interaction   64
    3.2.6 Effect of external stiffness   67
    3.2.7 Normal stress and lateral confinement   68
3.3 Laboratory testing on infilled rock joints   68
    3.3.1 Selection of infill material   70
    3.3.2 Preparation of infilled joint surface   70
    3.3.3 Setting-up the specimen in the shear boxes   71
    3.3.4 Application of normal load   71
    3.3.5 Shear behaviour of Type I joints   74
    3.3.6 Shear behaviour of Type II joints   77
    3.3.7 Effect of infill thickness on horizontal displacement corre-
           sponding to peak shear stress   77
    3.3.8 Effect of infill thickness on stress-path behaviour   81
    3.3.9 Effect of infill thickness on peak shear stress   83
    3.3.10 Drop in peak shear strength   83
    3.3.11 Strength envelope   87
3.4 Shear strength model for infilled joints   87
3.5 Remarks on infilled joint behaviour   90
References   91

4 MODELLING THE SHEAR BEHAVIOUR OF ROCK JOINTS   93
4.1 Introduction   93

4.2 Existing models based on CNS concept 94
    4.2.1 Model based on energy balance principles 94
    4.2.2 Mechanistically based model 97
    4.2.3 Graphical model 98
    4.2.4 Analytical model 98
4.3 Requirement of a new model 102
4.4 New shear strength model for soft rock joints 102
    4.4.1 Application of Fourier transform method for predicting unfilled joint dilation 102
    4.4.2 Prediction of normal stress with horizontal displacement 104
    4.4.3 Prediction of shear stress with horizontal displacement 105
4.5 Effect of infill on the shear strength of joint 113
    4.5.1 Hyberbolic modelling of strength drop associated with infill thickness 113
    4.5.2 Shear strength relationship between unfilled and infilled joints 115
    4.5.3 Determination of hyperbolic constants 115
4.6 Development of a computer code 119
4.7 Comparison between predicted and experimental results 119
    4.7.1 Dilation 119
    4.7.2 Normal stress 119
    4.7.3 Shear stress 124
    4.7.4 Strength envelopes 126
    4.7.5 Infilled joint strength 127
4.8 UDEC analysis of shear behaviour of joints 127
    4.8.1 Choice of joint models 127
    4.8.2 Continuous yielding model 131
    4.8.3 Conceptual CNS shear model 132
    4.8.4 CNL direct shear model 133
    4.8.5 Discretisation of blocks and applied boundary conditions 133
    4.8.6 Results and discussions 135
4.9 Summary of shear strength modelling 137
References 138

5 SIMPLIFIED APPROACH FOR USING CNS TECHNIQUE IN PRACTICE 140
5.1 Introduction 140
5.2 Underground roadway in jointed rock 140
    5.2.1 Boundary conditions 140
    5.2.2 Roadway excavation 141
5.3 Stability analysis of slope 143
    5.3.1 Limit equilibrium analysis (initial condition without bolts) 145
    5.3.2 CNS analysis (considering bolt contribution) 146
    5.3.3 CNL analysis considering joint contribution 147

6 HIGHLIGHTS OF ROCK JOINT BEHAVIOUR UNDER CNL AND
CNS CONDITIONS, AND RECOMMENDATIONS FOR THE FUTURE 149
  6.1 Summary 149
      6.1.1 Behaviour of unfilled joints under various rates of shear dis-
           placements 150
      6.1.2 Behaviour of unfilled joints under CNL and CNS 150
      6.1.3 Shear behaviour of unfilled and natural joints under CNS 150
      6.1.4 Behaviour of unfilled joints under CNS condition 151
      6.1.5 New shear strength model by the authors 151
  6.2 Recommendations for further study 152
      6.2.1 Modifications to laboratory procedures 152
      6.2.2 Field mapping 152
      6.2.3 Effective stress approach 152
      6.2.4 Bolted joints 153
      6.2.5 Scale effects 153
      6.2.6 Extension in numerical modelling 154
  References 154

APPENDIX: PROGRAM CODE FOR SHEAR STRENGTH MODEL 155

SUBJECT INDEX 163

Preface

This book is the result of more than 5 years of research conducted under Dr Buddhima Indraratna at University of Wollongong, in the Faculty of Engineering. The City of Wollongong being the centre of New South Wales coal mining activity, interest in the area of rock joints and jointed rock research has been significant. Geomechanics laboratories of University of Wollongong have been upgraded continuously to accommodate various demands from industry for testing rock specimens from underground mines. The contents of the book include the background of the subject in relation to rock mass behaviour and the major research findings of the authors during the recent years. As the behaviour of rock mass is predominantly governed by the behaviour of rock discontinuities, this volume will be a useful reference for mining and geotechnical practitioners and researchers.

The authors are pleased to acknowledge the support received from various organisations, including Strata Control Technology, Wollongong (SCT) and several local mining companies. In particular, the financial support provided by SCT has made it possible to develop some of the triaxial and shear testing equipment currently housed at University of Wollongong. The data obtained through these items of equipment have provided the basis of some significant concepts discussed in this book, such as the Constant Normal Stiffness and the Stress Path approach applied to jointed rock specimens.

The completion of this volume would not have been possible if not for the support of national and international colleagues. A large part of the subject materials and research data presented in this book have been reviewed on various occasions by practicing engineers and academics. Their comments and criticisms have undoubtedly enhanced the value of this book. While it is not possible to name them all, the authors are particularly thankful to Dr Winton Gale (SCT), Prof. Raghu Singh and Dr Naj Aziz (University of Wollongong), Jon Nemcik (SCT), Prof. P.H.S.W Kulatilake (University of Arizona), Dr Chris Haberfield (Monash University), Prof. Peter Kaiser (Laurentian University), Dr Jayantha Kodikara (Victoria University of Technology) and Dr David Toll (University of Durham). The contributions made by several postgraduate students, namely Ranjith Pathegama, Asoka Herath and Ashitav Dey are acknowledged. The untiring efforts of Alan Grant and Ian Laird (Geotechnical Technicians) during equipment design and laboratory testing programs are gratefully appreciated.

The field of jointed rock engineering is vast and the scope for further research and development is immense. The requirement for design changes in jointed rock and the challenge for acquiring further knowledge will grow with technology and the increased need for sub-surface space. A future extension of this work will include the stabilisation of rock joints and hydro-mechanics of jointed rock dealing with the coupled gas-water flow through discontinuities.

Dr Buddhima Indraratna and Dr Asadul Haque
Division of Civil, Mining & Environmental Engineering
University of Wollongong
Northfields Avenue, Wollongong
NSW 2522, Australia

CHAPTER 1

Characterization of joint surface roughness

1.1 INTRODUCTION

The majority of current rock joint models are capable of predicting the shear behaviour of relatively simplified joint surfaces. Many of these models do not include the complex joint surface characteristics, the effects of infill properties, and the degradation behaviour of asperities. In the field, joints are often infilled with clay and silt, which can be fully or partially saturated. A shear strength model should also address the role of Constant Normal Stiffness (CNS) boundary condition, which is commonly encountered in the field, for instance in underground mining situations. Current numerical techniques mainly rely on the conventional Constant Normal Load (CNL) strength parameters which are often inappropriate for:
– Evaluating the stability analysis of bedded mine roofs,
– Estimation of side shear resistance of rock-socketed piles, and
– Stability analysis of jointed strata subject to potential toppling failures.
The shear behaviour of non-planar joints is significantly influenced by the surface properties of the joints, or in other words, by the joint surface roughness. The rougher the joint, the greater the shear strength. Therefore, a numerical description of the roughness of a rock fracture surface is essential to the estimation of its shear strength, dilatancy, and stiffness. There are several methods of estimation of joint surface roughness. It is beyond the scope of this book to describe all of them. However, to have an understanding of the wide variety of methods dedicated to correctly represent the joint surface characteristics, the published literature can be categorised into the following four major techniques:
1. Joint Roughness Coefficient (JRC) measurement,
2. Self-similar and self-affine fractal method,
3. Spectral and line scaling method, and
4. Continuous mathematical functions such as Fourier transforms.

1

1.2   ROUGHNESS MEASUREMENT TECHNIQUES

1.2.1   *JRC measurement*

The degree of roughness is a measure of the inherent surface unevenness or waviness of the discontinuity relative to its mean plane. Barton (1973) proposed a joint roughness coefficient (JRC) to describe the surface roughness, using a scale from zero to 20. Typical roughness profiles for the entire JRC range are shown in Figure 1.1. This index is measured by the direct profiling method for the joint, or by an indirect method of performing a tilt test on a rough joint, together with a Schmidt Hammer Index test and the tilt test on a sawn rock surface. The joint roughness is estimated by visual matching with standard profiles, which is often subjective, or by back calculation using laboratory shear strength value. There-

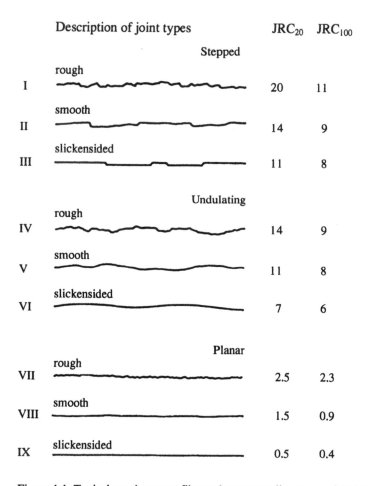

	Description of joint types	JRC_{20}	JRC_{100}
	Stepped		
	rough		
I		20	11
	smooth		
II		14	9
	slickensided		
III		11	8
	Undulating		
	rough		
IV		14	9
	smooth		
V		11	8
	slickensided		
VI		7	6
	Planar		
	rough		
VII		2.5	2.3
	smooth		
VIII		1.5	0.9
	slickensided		
IX		0.5	0.4

Figure 1.1. Typical roughness profiles and corresponding range of JRC (after Barton 1973).

fore, the use of such methods in numerical roughness calculations is not always viable.

Czichos (1978) described two methods of characterisation of surface roughness in his book 'Tribology Series, 1'. In the first method, a couple of parameters have been proposed in relation to a cross-section of the surface to define surface roughness (Fig. 1.2).

The most commonly used height parameter in surface profilometry is the R_a value or center line average value, which is the average deviation of the profile from the reference mean line. The other height parameter is the maximum peak to valley height, R_t within the sample length, l (Fig. 1.2).

The second method is to use a profilometer to describe the surface topography as an electrical signal and to analyse it statistically. Finally, the surface profile is defined in terms of the following functions:

– The probability distribution of ordinate height, which can be distributed in a Gaussian manner.
– The auto correlation function of the profile defined as:

$$Z(\Delta) = \lim_{L \to \infty} \frac{1}{L} \int_{-L/2}^{+L/2} y(x) y(x + \Delta) \, dx \tag{1.1}$$

where, $y(x)$ = height of a profile at a given coordinate x, and $y(x + \Delta)$ = height at an adjacent coordinate $(x + \Delta)$

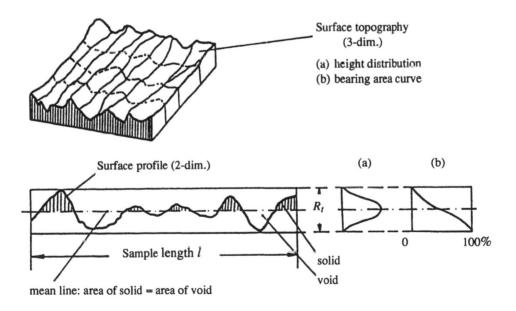

Surface topography (3-dim.)

(a) height distribution
(b) bearing area curve

Figure 1.2. Characteristics of surface roughness (after Czichos 1978).

The method suggested by Xie & Pariseau (1992) can be used to define the value of JRC for regular saw-teeth profiles in terms of fractal dimensions, as explained below:

$$JRC = 85.2671 \times (D-1)^{0.5679} \tag{1.2}$$

where

$$D = \frac{\log(4)}{\log\left[2\left(1 + \cos\tan^{-1}\left(\frac{2h}{l}\right)\right)\right]}$$

In the above D = fractal dimension, h = average height of asperity, and L = average base length of asperities.

1.2.2 *Fractal method*

Seidel & Haberfield (1995) formulated a fractal method to characterise the joint surface roughness based on fractal geometry, fractal dimension and self-similarity. The fractal dimension and standard deviations of angle and height of a joint profile of unit direct length (Fig. 1.3) can be characterised as:

$$s_\theta \approx \cos^{-1}(N^{(1-D)/D}) \tag{1.3}$$

$$s_h \approx \sqrt{N^{-2/D} - N^{-2}} \tag{1.4}$$

$$D \approx -\frac{\log(N)}{\log\left(s_h^2 + \frac{1}{N^2}\right)^{1/2}} \tag{1.5}$$

where D = fractal dimension, s_θ and s_h = standard deviation to angle and height, N = number of segments.

Figure 1.3. a) Single chord geometry, and b) Definition of standard deviation of chord length (after Seidel & Haberfield 1995).

For a line of direct length, L_d, the standard deviation of height is given as:

$$s_h \approx L_d \sqrt{N^{-2/D} - N^{-2}} \qquad (1.6)$$

Using the mid-point displacement technique, Seidel & Haberfield (1995) established the standard deviation of height for kth bisection as follows:

$$s_{h.k} \sim \sqrt{N^{-2(1=kD-D/D)} - N^{-2k}} \qquad (1.7)$$

The standard deviation of angle can be determined as follows:

$$s_{\theta.k} = \sqrt{k}\,s_{\theta.1} \approx \sqrt{k}\,\cos^{-1}\left(2^{(1-D)/D}\right) \qquad (1.8)$$

Examples of typical profiles obtained by applying the mid-point displacement technique described above are shown in Figure 1.4.

The surface roughness of joints plays an important role on the strength and deformation behaviour of natural single joints. Kulatilake et al. (1998) developed a new peak shear strength criterion which incorporates the fractal joint parameters. The fractal parameters was estimated to quantify natural rock joint roughness using the Roughness-length method. For a self-affine profile, the following relationship can be written (Malinverno 1990):

$$\log_e [s(w)] = \log_e A + H \log_e w \qquad (1.9)$$

where $s(w)$, w, H and A are respectively, the standard deviation of the profile height, the spanning length of the profile, the Hurst exponent and a proportionality constant. The parameters H ($H = 2 - D$, $D =$ fractal dimension) and A can be

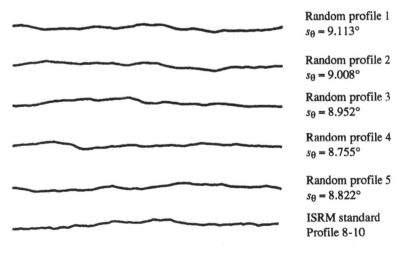

Figure 1.4. Random roughness generation for s_θ=9° compared with ISRM standard roughness profile JRC 8-10 (after Seidel & Haberfield 1995).

estimated from the slope and intercept of the plot of ln $s(w)$ and ln w, respectively. The value $s(w)$ is calculated as the root-mean-square (RMS) value of the profile height residuals on a linear trend fitted to the sample points in a window of length w. The estimation of a trend from the data can be made by calculating the RMS roughness using m_i-2 degrees of freedom as:

$$s(w) = \text{RMS}(w) = \frac{1}{n_w} \sum_{i=1}^{n_w} \sqrt{\frac{1}{m_i-2} \sum_{j=w_i} (z_i - \bar{z})^2} \qquad (1.10)$$

where n_w = total number of windows of length w, m_i = number of points in window w_i, z_i = residuals on the trend, and $\bar{z}$ = mean residual in window w_i.

In order to obtain a reliable estimates of fractal parameters of natural rock joint profile, Kulatilake et al. (1998) suggested a data density, d, greater than or equal to about 5.1 for the profile. For roughness profiles having $5.1 \le d \le 51.2$ and $D \le 1.70$, w values between 2.5% and 10% of the profile length are highly suitable to produce accurate estimates of fractal parameters.

The two fractal parameters (D, A) were used to quantify the stationary roughness of model replicas of natural joints. To obtain these parameters, the joint surface was first digitized using a profilometer. For a given set of values of d and w, the average fractal dimension and spanning length can be computed from the following relationships:

$$\text{Average } D = \frac{\sum_{j=1}^{N} (D)_j L_j}{\sum_{j=1}^{N} L_j} \qquad (1.11a)$$

$$\text{Average } A = \frac{\sum_{j=1}^{N} (A)_j L_j}{\sum_{j=1}^{N} L_j} \qquad (1.11b)$$

where, $(D)_j$ = estimated fractal dimension for jth surface height, $(A)_j$ = proportionality constant for jth surface height profile, N = number of profiles, and L_j = straight length of the jth surface height profile in the direction considered.

Finally, the above roughness parameters were used to establish a new shear strength criterion of rock joint in the following form:

$$\frac{\tau}{\sigma_j} = \frac{\sigma_n}{\sigma_j}(1-a_s)\tan\left[\phi_b + \left(160D^{5.63}A^{0.88} \pm 1.8I_{eff}\right)\log\left(\frac{\sigma_j}{10\sigma_n}\right)\right]$$
$$+ a_s\frac{\tau_r}{\sigma_j} \qquad (1.12)$$

where a_s = area of asperity shear, I_{eff} = effective non-stationary trend angle, τ = shear strength, σ_j = joint compressive strength, and σ_n = normal stress.

Geostatistical technique introduced by Piggott & Elsworth (1995) describes both surface roughness and the spatial correlation of elevation. A self-affine fractal is a feature, $z(x)$, which exhibits scaling behaviour in the form of:

$$z(sx) \approx s^H z(x) \qquad (1.13)$$

where s = a scale factor, x = a coordinate vector, and H = related to fractal dimension (D) by

$$D = E + 1 - H \qquad (1.14)$$

where E = Euclidean dimension of the feature ($E = 2$ for a fracture surface and $E = 1$ for a section through the surface).

A self-affine fractal surface has a semi-variogram function of surface elevation of the form:

$$Y(h) \approx |h|^\alpha \qquad (1.15)$$

where h = lag vector.

The semi-variogram functions of surface elevation of the profiles were calculated using,

$$Y(h_i) = \frac{1}{2(n-1)}\sum_{j=1}^{n-i}(z_{i+j} - z_j)^2 \qquad (1.16)$$

Figure 1.5 represents the semi-variogram functions of profiles AECLY and AECLX. Results show a discrepancy between calculated results and theoretical semi-variogram functions for increasing lag distance.

McWilliams et al. (1990) introduced a modified divider method for the fractal calculations to characterize rock fracture roughness. In this method, the profile length (L) is related to the fractal dimensions according to the following relationship:

$$\log(L) = I + (1 - D)\log(r) \qquad (1.17)$$

where r = divider span, L = estimated profile length, D = fractal dimension = $1 - \beta$, and I = vertical intercept on log scale. A graphical procedure is shown in Figure 1.6 outlining the modified divider method for fractal calculations.

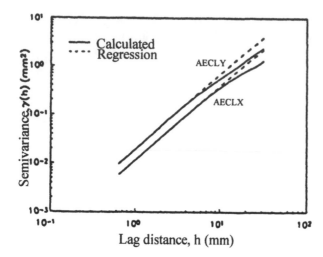

Figure 1.5. Semi-variogram functions of surface elevation (after Piggott & Elsworth 1995).

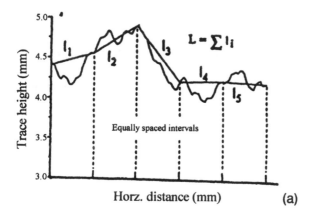

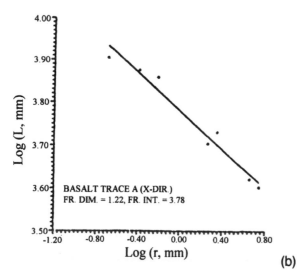

Figure 1.6. Modified divider method to estimate fractal dimension: a) Increments used to obtain L, and b) Plot of log (L) vs log (r) (after McWilliams et al. 1990).

1.2.3 *Spectral method*

Durham & Bonner (1995) proposed a spectral method to estimate surface roughness of rock joints. Firstly, the rock surface is digitized using a profilometer by recording coordinates (x, y, z) at a given point. The power spectral density (PSD) is then calculated for each x-z profile taken, and then averaged to produce a single estimate for the entire surface. The PSD is calculated based on the direct method of classical spectral estimation:

$$G_i(f) = \frac{h^2}{L} |Z_i(f)|^2 \qquad (1.18)$$

where h = sampling interval, L = length of the profile, and $Z_i(f)$ = fast Fourier transformation (FFT) of the discretely sampled profile.

Using Equation 1.18, Durham & Bonner (1995) have calculated PSD for three samples based on extensively digitized data at sampling intervals from 0.05 to 0.5 mm and profile length of 65 mm. A typical PSD plot is shown in Figure 1.7. To cover as much as surface area as possible, they have digitized each surface broadly at coarse scale and smaller patches at a finer scale. As expected, the fine and coarse scale PSD curves overlap usually, and the single and different topographies also bear a logical relationship. At low spatial frequencies, single surface topographies continue to gain power and difference topographies lack power.

Spectral Analysis as outlined by Piggott & Elsworth (1995) describes both the roughness and spatial correlation of fracture surface topography. The spectral method uses the spectral density function of surface elevation as:

$$\Gamma(f) \approx f^{-\beta} \qquad (1.19)$$

where f = spatial frequency.

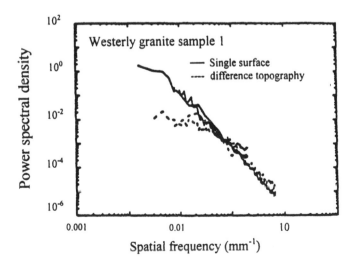

Figure 1.7. PSD plot for Westerly granite sample (after Durham & Bonner 1995).

The semi-variogram function can be obtained from spectral density function using the Fourier transform relation as:

$$Y(h) \approx \int_0^\infty \Gamma(f) \left[1 - \cos(2\pi f h)\right] df \tag{1.20}$$

Substituting wavelength, $\lambda = f^{-1}$ and $\Gamma(f)$ in Equation 1.20, the following function can be obtained for $Y(h)$ as:

$$Y(h) \approx h^{\beta-1} \int_0^\infty \lambda_d^{\beta-2} \left[1 - \cos\left(\frac{2\pi}{\lambda_d}\right)\right] d\lambda_d \tag{1.21}$$

Finally, the fractal dimension can be represented as:

$$D = \frac{2E + 3 - \beta}{2} \tag{1.22}$$

The spectral density functions of the profiles were calculated using the discrete Fourier transform relation as:

$$\Gamma(\lambda_i) \approx \left[\frac{1}{n}\sum_{j=1}^n w_j z_j \cos\left(\frac{2\pi ij}{n}\right)\right]^2 + \left[\frac{1}{n}\sum_{j=1}^n w_j z_j \sin\left(\frac{2\pi ij}{n}\right)\right]^2 \tag{1.23}$$

The calculated results for the simulated and measured profiles are shown in Figure 1.8 based on 1024 and 256 point segments, respectively.

Other method that need to be mentioned is the statistical analysis by Piggott & Elsworth (1995). This technique is used to describe the overall roughness of the surface and distribution of elevations. This method does not describe the spatial correlation of surface elevation and thereby, cannot be used to measure fractal dimension. The mean (μ) and variance (σ^2) of surface are given by:

$$\mu = \int_{-\infty}^\infty z p(z) dz \approx \frac{1}{n}\sum_{i=1}^n z_i \tag{1.24}$$

and

$$\sigma^2 = \int_{-\infty}^\infty (z - \mu)^2 p(z) dz \approx \frac{1}{n}\sum_{-\infty}^\infty (z_i - \mu)^2 \tag{1.25}$$

where $p(z)$ = probability distribution function of surface elevation.

The standard deviation of surface elevation, σ, is the root-mean-square (RMS) roughness of the surface. Equations 1.24 and 1.25 are applied to AECLX and AECLY profiles in Figure 1.9, which subsequently yield values of σ of 1.02 and

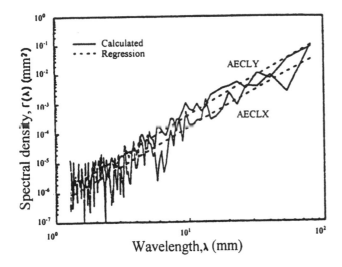

Figure 1.8. Spectral density functions of surface elevation (after Piggott & Elsworth 1995).

1.94 mm, respectively. The distribution of elevation magnitudes can be represented by $p(z)$ as:

$$p(z) = \sum_{-\infty}^{\infty} p(z)dz \qquad (1.26)$$

For discrete elevation data sorted into ascending order, Equation 1.26 may be approximated as:

$$p(z_i) = \frac{2i-1}{2n} \qquad (1.27)$$

A plot of $p(z)$ with elevation, z (Fig. 1.10) for those two profiles shows a normal population distribution model and fits a linear regression model having σ values closer to Equation 1.25. This suggests that the distribution of fracture surface elevation may be represented by a normal population distribution model.

1.2.4 *Digital coordinate measuring machine*

In order to assess the roughness of the joint profiles, an interface can be examined under a digital Ferranti (Mercury) Coordinate Measuring Machine (CMM) shown in Figure 1.11. The CMM is manually driven, which consists of a set of Renishaw probes and a MICRO 900 microprocessor. The basic frame of the machine is placed on a granite table. The machine can measure a minimum of 1 micron position resolution and can achieve an accuracy with 95% confidence under normal working conditions.

The test specimens are required to place on the granite table of the CMM. Be-

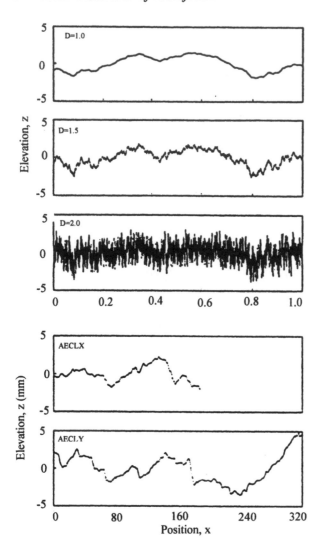

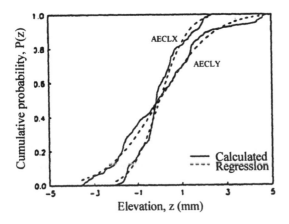

Figure 1.9. Simulated ($D = 1.0$, 1.5 and 2.0) and measured (AECLX and AECLY) fracture surface profiles (after Piggott & Elsworth 1995).

Figure 1.10. Cumulative probability distribution functions of surface elevation (after Piggott & Elsworth 1995).

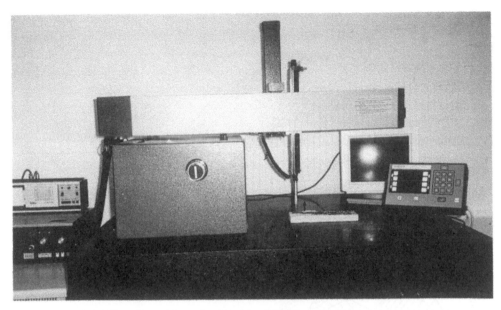

Figure 1.11. Digital Coordinate Measuring Machine (CMM).

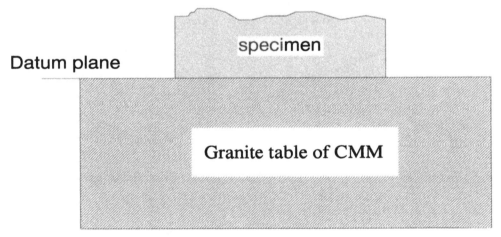

Figure 1.12. Datum surface of CMM (modifed after Islam 1990).

fore measuring the coordinates of the specimen, a datum plane is required to be established with respect to which the measurements were recorded. For simplicity, the surface of the granite table can be considered as a 'perfect datum' as shown in Figure 1.12.

The surface profile of the specimen is examined using the CMM, and coordinates at many points are recorded using a touch trigger probe. All the measurements should be taken with respect to the 'perfect datum'. The surface profiles of

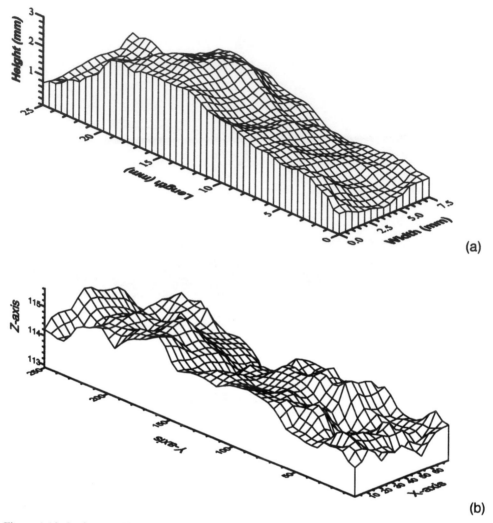

Figure 1.13. Surface profiles of: a) Model joint, and (b) Field specimens obtained from CMM.

the natural (tension) joints and natural (field) specimens collected from Kangaroo Valley site of New South Wales are shown in Figure 1.13. The average height of the joint asperities varied from 1.22 to 2.35 mm.

1.2.5 *Fourier transform method and its application to measure surface roughness*

Fourier Transform method can be used precisely to characterize the surface roughness of joints. The measured variations in asperity height corresponding to a given interval of sample length are required as input parameters. The corresponding Fourier coefficients are then obtained. The dilation of the joints at any shear displacement can be predicted using Equation 1.28.

$$d_v(h) = \frac{a_0}{2} + \sum_{n=1}^{\infty} \left[a_n \cos\left(\frac{2\pi n h}{T}\right) + b_n \sin\left(\frac{2\pi n h}{T}\right) \right]$$ (1.28)

where $d_v(h)$ = dilation at any displacement, h; a_0, a_n, b_n = Fourier coefficients; T = period and n = number of harmonics. The Fourier coefficients for the prediction of dilation of a joint in the field can be obtained from a series of laboratory tests conducted on field joints collected from the same site. A database which stores the likely values of Fourier coefficients under different normal stresses for a particular site can be used for the prediction of shear behaviour of joints. The usage of Fourier transform method to characterise a particular joint surface will be discussed in detail in Chapter 4.

1.3 PHYSICAL AND MECHANICAL PROPERTIES OF INTACT AND JOINTED ROCKS

The physical and mechanical properties of rocks are very important in investigation and design of structures in intact or fractured rock mass. The physical properties of rocks (e.g. density, porosity, water content) have significant effect on the strength and deformation characteristics. However, it is not the scope of this book to explain them in depth. The reader should refer to fundamental rock mechanics text book for these details. Hudson (1988) and Dershowitz & Einstein (1988) reviewed the variables required for quantitative description of rock joints. In this book, only the factors which control the shear strength of jointed rock mass are discussed briefly.
– Friction angle of planar joints,
– Joint water pressure,
– Compressive strength,
– Surface geometry, and
– Presence of infill material.
The friction angle has direct effect on the shear strength of rock joints. The shear strength of joints increases with the increase in friction angle. However, the presence of water decreases the total normal stress applied to the interface, thereby, reducing the shear strength of joints. The joint surface geometry has significant effect on the shear behaviour of rough joints as demonstrated by Patton (1966). The mechanical properties of the joints are to a great extent dependent upon whether the joints are clean and closed, or open and filled with some infill material. The infill material may sometimes act as a cement bond (sealant) to the joints, and in such cases, it is rarely regarded as a joint. In other cases, the infill material may consist of partially loose to completely loose cohesionless soils (e.g. sand, coarse fragmentary material etc.) which are deposited in open joints be-

tween the two surfaces of the joints. It may also be produced as a result of weathering and decomposition of the joint wall itself. Sometimes, the strength of the infilled joint is considered smaller than that of the infill material as observed by Kanji (1974). However, for infill thickness smaller than twice the asperity height, the interaction between the asperities and the infill material will influence the shear behaviour. The effect of surface roughness and infill on the joint shear strength will be explained in detail in subsequent chapters.

REFERENCES

Barton, N. 1973. Review of a new shear strength criteria for rock joints. *Engineering Geology*, 7: 287-332.

Czichos, H. 1978. *Tribology – a systems approach to the science and technology of friction, lubrication and wear.* Elsevier Scientific Publishing Company, Amsterdam, 400 p.

Dershowitz, W.S. & Einstein, H.h. 1988. Characterizing rock joint geometry with joint system models. *Rock Mech. and Rock Engng*, 21: 21-25.

Durham, W.B. & Bonner, B.P. 1995. Closure and fluid flow in discrete fractures. *Fractured and Jointed Rock Masses*, Balkema Publishers, Rotterdam, pp. 441-446.

Hudson, J.A. 1988. The role of rock characterization in slope stability analysis. *Proc. Secondo Conference di Mechanica e Ingegneria delle Rocce, COREP-Polytechnico di Torino.*

Islam, M.N. 1990. An analysis of machining accuracies in CNC machining operation. M.E (Hons) thesis, University of Wollongong, Australia.

Kanji, M.A. 1974. Unconventional laboratory tests for the determination of the shear strength of soil-rock contacts. *Proc. 3rd Congr. Int. Soc. Rock Mech., Denver*, 2: 241-247.

Kulatilake, P.H.W.S., Um, J., Panda, B.B. & Nghiem, N. 1998. Accurate quantification of joint roughness and development of a new peak shear strength criterion for anistropic rock joints. *Proc. Int. Conf. on Geomechanics/Ground Control in Mining and Underground Construction, Wollongong, Australia*, 1: 33-48.

Malinverno, A. 1990. A simple method to estimate the fractal dimension of a self-affine series. *Geophysical Research Letters*, 17: 1953-1956.

McWilliams, P.C., Kerkering, J.C. & Miller, S.M. 1990. Fractal characterization of rock fracture roughness for estimating shear strength. *Mechanics of Jointed and Faulted Rock*, Balkema Publishers, Rotterdam, pp. 331-336.

Patton, F.D. 1966. Multiple modes of shear failure in rocks. *Proc. 1st Cong. ISRM, Lisbon*, pp. 509-513.

Piggott, A.R & Elsworth, D. 1995. A comparison of methods of characterizing fracture surface roughness. *Fractured and Jointed Rock Masses*, Balkema Publishers, Rotterdam, pp. 471-476.

Seidel, J.P. & Haberfield, C.M. 1995. Towards an understanding of joint roughness. *Rock Mechanics & Rock Engineering*, 28(2): 69-92.

Xie, H. & Pariseau, W.G. 1992. Fractal estimation of joint roughness coefficients. *Proc. Int. Conf. on Fractured and Jointed Rock Masses*, Balkema Publishers, Rotterdam, pp. 125-131.

CHAPTER 2

Shear behaviour of clean rock joints

2.1 INTRODUCTION

The presence of joints in a rock mass has a significant influence on its shear strength and deformation characteristics. In hard rocks, for relatively closely spaced joints, for example 1 m spacing, the mechanical behaviour of the rock mass is similar to that of the joints in relation to the strength of the intact rock material (Lama 1978).

In the past, various research projects have been conducted on rock joints in the laboratory using conventional direct shear apparatus, where the normal stress acting on the joint interface is considered to be constant throughout the shearing process. Therefore, this particular mode of shearing is suitable for planar joints, where the joint does not dilate during testing, hence, the normal stress remains constant during shearing.

However, for non-planar joints, dilation results as shearing progresses, and if the surrounding rock mass inhibits some of this dilation, then an inevitable increase in normal stress occurs. Therefore, the shearing of rough joints no longer takes place under constant normal load (CNL); it is the stiffness of the surrounding rock mass that controls the shear behaviour. This condition is defined as shearing under Constant Normal Stiffness (CNS).

In the last two decades, researchers have tried to explain the behaviour of rough rock joints under CNS either through laboratory testing, or through analytical methods.

Recent studies show that CNS strength parameters are more representative for the design of rock socketed piles and underground excavations (Johnston et al. 1987).

In this chapter, studies conducted on the behaviour of clean/unfilled rock joints are discussed, particularly those of experimental modeling. The behaviour of filled joints is discussed in Chapter 3. For discussion purposes, the behaviour of the unfilled joints under CNS is classified into the following categories:
− Tests on hard joints,
− Tests on soft joints, and
− Modelling of joint shear behaviour.

2.2   CNS AND CNL CONCEPTS

The presence of joints in a rock mass can affect its mechanical behaviour, depending on the underground situation. When dilation of the rock joints during shearing is constrained or partially constrained, an increase in the normal stress over the shear plane occurs, which substantially increases the shear resistance. Figure 2.1 shows an underground excavation where potentially unstable rock blocks are constrained between two parallel dilatant rock joints. The sliding of such blocks inevitably increases the normal stress, and also, dilation becomes significant if the joint surfaces are rough. The increase in normal stress on the shear plane is equal to $k_n.\delta_v$, where k_n is the normal stiffness of the surrounding rock mass and δ_v is the dilation. Tests conducted under Constant Normal Load (CNL) condition yield shear strengths that are too low for such practical situations (Goodman 1976).

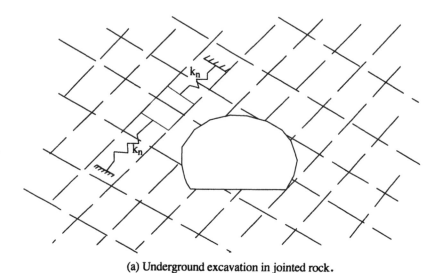

(a) Underground excavation in jointed rock.

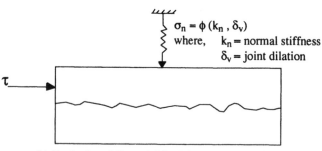

(b) Equivalent 2-D model for joint on the top of roof.

Figure 2.1. Joint behaviour on the roof of an excavation (after Indraratna et al. 1999).

As another example, Figure 2.2 shows a rock socketed pile where the interface between the concrete and the socket is considered to be rough. When this pile is loaded vertically, the side shear resistance develops as a function of the variable normal stress associated with the dilation of the rough joint surface. The deformation mechanism and the simplified 2-D models are illustrated in Figures 2.2b-d.

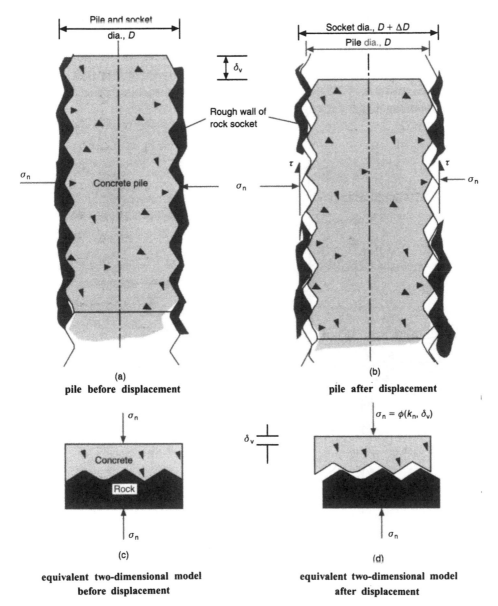

Figure 2.2. Idealised displacement behaviour of pile socketed in rock (after Johnston & Lam 1989, Indraratna et al. 1999).

Indraratna (1987) pointed out that the shearing process of a bolted joint also takes place under CNS condition. Details of deformation behaviour of a bolted joint during shearing is shown in Figure 2.3.

Another situation where the normal load changes during shearing is the earthquake shaking of rock slopes, where the direction of shearing and the magnitude of normal load on any potential sliding plane are variable during shaking. However, dynamic effects are not considered within the scope of this chapter.

In general, the CNL condition is only realistic for shearing of planar interfaces, where the normal stress applied to the shear plane remains relatively constant such as in the case of rock slope stability problems. However, for situations as illustrated in Figures 2.1-2.3, the development of shear resistance is a function of constant normal stiffness (CNS), and the use of CNL test results for such cases leads to underestimated shear strengths.

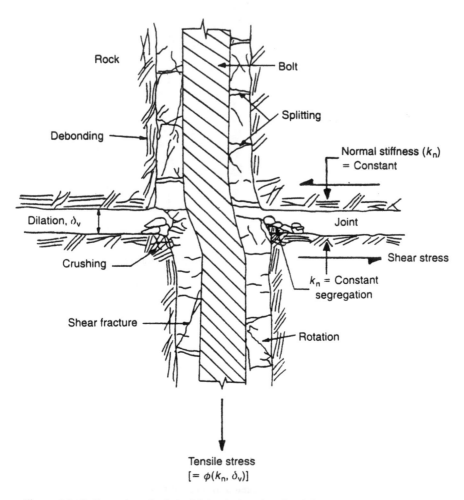

Figure 2.3. Deformation of a bolted joint during shearing (after Indraratna et al. 1999).

2.3   SHEAR APPARATUS FOR LABORATORY TESTING

The importance of conducting CNS testing has been discussed in Sections 2.1 and 2.2. For the last two decades, researchers have been involved in the modified design of direct shear apparatus for a wide range of applications, where the accuracy of shear strength parameters have been considered to be important (e.g. Benmokrane & Ballivy 1989, Skinas et al. 1990, Ohnishi & Dharmaratne 1990). In the mid 80's, Monash University (Melbourne) designed a direct shear apparatus for finding the side shear resistance of rock socketed piles under CNS condition (Johnston et al. 1987). This apparatus can also be used for conventional shear testing of joints, i.e. under CNL condition.

In order to investigate the shear behaviour of soft rock/rock interface and infilled joints, Indraratna et al. (1997) designed a large-scale direct shear apparatus at University of Wollongong, Australia. This apparatus can be used to test specimens under both CNL and CNS conditions. This direct shear apparatus consists of a pair of large shear boxes, the shear and normal loading devices, and displacement monitoring transducers. A brief summary of different components are given below.

2.3.1   *Large scale shear boxes*

The CNS shear apparatus consists of two steel boxes, one of size 250 mm in length, 75 mm in width and 150 mm in height at the top, and the other of size $250 \times 75 \times 100$ mm at the bottom. A set of springs are used to simulate the normal stiffness of the surrounding rock mass, $k_n = dN/d\delta_v$, where, dN and $d\delta_v$ are the changes in normal load and displacement, respectively. As shown in Figure 2.4, the top box can only move in the vertical direction along which the stiffness is kept constant (8.5 kN/mm). The bottom box is fixed on a rigid base through bearings, and can move only in the shear (horizontal) direction.

2.3.2   *Loading device*

The desired initial normal stress (σ_{no}) is applied by a hydraulic jack, where the applied load is measured via digital strain meters connected to a calibrated load cell. The maximum normal load capacity of the apparatus is 180 kN. The shear load is applied via a transverse hydraulic jack which is connected to a strain controlled unit (Fig. 2.4). The applied shear load can be recorded via strain meters fitted to a load cell. The apparatus has a maximum shear load capacity of 120 kN, and the rate of horizontal displacement can be varied between 0.35 and 1.70 mm/min. During the testing of specimens under CNL, an electric pump is connected to the hydraulic jack to ensure the application of a constant load on the shear plane.

Figure 2.4. A close view of the large-scale CNS direct shear apparatus (after Indraratna et al. 1998; reprinted with kind permission from Kluwer Academic Publishers)).

2.3.3 *Measurement of displacements*

The dilation and the shear displacement of the joint are recorded by dial gauges mounted on the center of the top specimen and in the horizontal (shear) direction, respectively (Fig. 2.4). The displacements can be read with an accuracy of 0.01 mm/division. The dial gauge mounted on the top is also used to read the consolidation settlements of the infilled joint under the applied initial normal stress before shearing of the joint.

2.4   LABORATORY MODELLING OF ROCK JOINTS

2.4.1   *Selection of model material for joint*

Gypsum plaster ($CaSO_4.H_2O$ hemihydrate, 98%) is widely used to make idealised soft rock joints, mainly because this material is universally available and is inexpensive. It can be moulded into any shape when mixed with water, and the long term strength is independent of time once the chemical hydration is complete. The initial setting time of plaster is about 25 minutes when mixed with 60% water by weight. The basic properties of the model material are determined by performing many tests on 50 mm diameter specimens after a curing period of two weeks, at an oven-controlled temperature of 50°C. The cured plaster specimen shows a consistent uniaxial compressive strength (σ_c) in the range of 11 to 13 MPa and a Young's modulus (E) of 1.9 to 2.3 GPa. It is found to be suitable for simulating the behaviour of a number of jointed soft rocks such as coal, friable limestone, clay shale and mudstone (Indraratna 1990). A comprehensive evaluation of the gypsum plaster rock based on dimensionless strength factors is given elsewhere by Indraratna (1990).

2.4.2   *Preparation of saw-tooth and natural specimens*

The preparation of model joints depends on the shear box assembly, type of joints and joint surface roughness. In this book, the procedure followed by the authors to prepare regular triangular soft joints and replica of field joints are discussed in relation to the CNS direct shear apparatus (Indraratna et al. 1998). The top and bottom moulds of the CNS shear apparatus are detached before casting the upper and lower portions of the specimen inside them. Gypsum plaster is initially mixed with water in the ratio of 5:3 by weight. Subsequently, the bottom mould together with the adjustable collar infilled with the mixture, and left for at least an hour to ensure adequate hardening before casting the upper specimen. Naturally, the collar is shaped according to the desired surface profile, in accordance with triangular asperities with angles of inclination of 9.5° (Type I), 18.5° (Type II) and 26.5° (Type III). Natural (tension) joint profiles are obtained from splitting a sandstone specimen under Brazilian test. In order to cast identical specimens of natural (tension) joints, the joint profiles are reproduced from sandstone surfaces using a resin based compound, commercially known as 'plastibond'. Plastibond is initially spread over the sandstone surface and left for 24 hours to harden. The reproduced surface is then cut according to the size of the shear box. After one joint profile is cast, the top mould is then placed over the bottom mould and filled with the plaster mixture. A thin polythene sheet is inserted between the two moulds separating the two fully mated joint surfaces, and the whole assembly is subsequently cured for another hour at room temperature to complete initial setting. During specimen preparation, mild vibration can be applied to the moulds externally to eliminate any entrapped air. Once initial hardening has taken place, the moulds are stripped and the specimens are cured at 50°C inside an oven for two weeks. Before testing, the specimens are allowed to reach the room temperature.

2.4.3 *Sampling and preparation of field joint specimens*

The natural joints are sampled from a rockslide site at Kangaroo Valley in New South Wales, Australia (Fig. 2.5). The Hawkesbury Sandstone forms the cliffs around the Kangaroo Valley, providing numerous waterfalls and several mesas such as Broughton Head. The average thickness of sandstone in the vicinity of the Kangaroo Valley is approximately 120 m. The Hawkesbury Sandstone overlaps the Narrabeen Group and overlies the Illawarra Coal Measures to the west of Fitzroy Falls. Two main types of mudrock interbeds are seen in the Hawkesbury Sandstone, namely dark-grey claystones and mid-grey siltstone/fine sandstone laminates. Generally, the dark-grey claystone infill washouts while the laminates grade into the underlying sandstone. Jointing is prominent in the Hawkesbury Sandstone (Fig. 2.6). Petrological studies show that it is a poorly sorted medium to coarse grained sandstone having 68-70% quartz (Geological Survey of New South Wales 1974).

The field joints are cut at the site as a block and then transported to the labo-

Figure 2.5. Location of the rockslide at Kangaroo Valley of New South Wales.

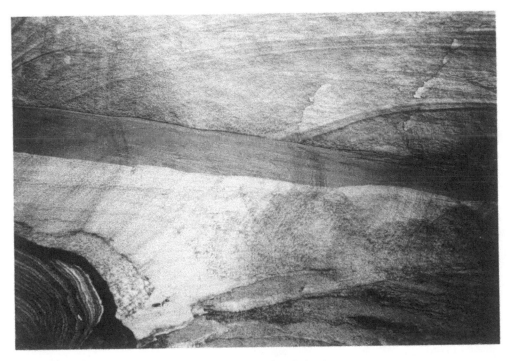

Figure 2.6. Orientation of joint plane in the field (sectional view).

ratory. The laboratory specimens are prepared by cutting the joint to fit the top part of the mould of size of $250 \times 75 \times 150$ mm, and the bottom part of mould of size of $250 \times 75 \times 100$ mm. A close view of the prepared joint is shown in Figure 2.7.

The joints are composed of highly weathered sandstone (Hawkesbury sandstone) having an uniaxial compression strength (σ_c) of 19-21 MPa, as measured by the authors in the laboratory. Pells & Turner (1978) reported an uniaxial strength in the range of 20-33 MPa for the Hawkesbury Sandstone in the Sydney region. A basic friction angle (ϕ_b) of 32° is observed for saw-cut surfaces of the collected joints.

2.5 LABORATORY TESTING OF ROCK JOINTS

In the past, several researchers have conducted tests on both artificial and natural unfilled/clean joints to explain the shear behaviour under CNS. Fundamental research work has been carried out initially by Patton (1966), and subsequently by Ladanyi & Archambault (1970), Barton (1973, 1976, 1986), Hoek (1977, 1983, 1990), Hoek & Brown (1980), Bandis et al. (1981), Hencher & Richards (1989), Kulatilake (1992), Kulatilake et al. (1995), Saeb & Amadei (1992), Brady &

Figure 2.7. A close view of natural (field) joint prepared for testing.

Brown (1985, 1993) among others. In all these studies, the CNL or zero normal stiffness condition is considered. It is not within the scope of this book to present a critical review of all such CNL testing results, as the aim of the authors is to focus more on the importance of CNS testing. The aspects of CNS testing are critically discussed in the following sections of this Chapter, in relation to the overall shear behaviour of joints, shear displacement rate, change in normal stress and dilation behaviour.

2.5.1 *Shear strength response under Constant Normal Stiffness*

The shear strength of non-planar joint increases due to the application of external normal stiffness, k_n, which allows a joint to shear under restricted dilation. It has been reported that the peak shear stress of unfilled joints increases under CNS in comparison to that of the CNL (Ohnishi & Dharmaratne 1990, Obert et al. 1976, Indraratna et al. 1998, Skinas et al. 1990). Lam & Johnston (1989) demonstrated that CNS method can closely model the shear behaviour of rock socketed piles by conducting tests on concrete/rock interfaces. Ohinishi & Dharmaratne (1990) reported that the shear strength obtained under CNS is greater than that of CNL for a medium strength cement mortar joint. Moreover, the shear strength is observed to increase with the increase in initial normal stress (σ_{no}).

Benmokrane & Ballivy (1989) reported that CNS has a large influence on the shear strength of rock joints and cement mortar joints having rough interfaces and higher strength (σ_c = 90 MPa). As the dilation occurs for rough joints under CNS condition, the normal stress acting on the interface increases, thus contributing to an increased value of shear stress. However, the CNS method has no influence on the behaviour of flat joints which do not produce any dilation during shearing. Kodikara & Johnston (1994) reported test results on irregular and regular concrete/soft rock joints and concluded that regular joints have a higher shear resistance than irregular joints. The mechanism of failure is very similar under both CNS and CNL conditions. In contrast, the CNS shear behaviour is almost identical to that of CNL under very high normal stress, where all the asperities undergo shearing without significant dilation (Leichnitz 1985).

Haberfield & Seidel (1998) extended the CNS technique on soft rock/rock joints and found that the failure mechanism of rock/rock joints is significantly different from concrete/rock joints. For concrete/rock joints, the much stronger concrete part of the joint constrained the failure over the entire contact length of each asperity. However, for rock/rock joints, the material on both sides of the interface is similar, allowing failure to occur at localised regions of high stress. Failure gradually progresses until complete failure of each asperity occurs and this results in a significant reduction in strength. Failure of soft rock/rock joints occurs on a curved surface, resulting in the development of a sliding/shearing mechanism. In contrast, tests on sandstone/sandstone joints reveal that a wearing mechanism is the dominant form of failure. Test results obtained by Van Sint Jan (1990) for soft model joints (σ_c = 0.92 MPa) reveals that an increase in stiffness increases the shear strength of joints and that the peak shear stress is always attained before the development of maximum normal stress.

The effect of CNL and CNS conditions on the strength envelope is assessed by plotting the shear strength against the corresponding normal stress. It has been observed that the peak friction angle obtained under CNS is always smaller than that of the CNL condition (Ohinishi & Dharmaratne 1990, Van Sint Jan 1990).

2.5.2 *Normal stress and dilation behaviour*

Joint dilation decreases under CNS condition, ie. the greater the normal stiffness surrounding the shear plane, the smaller the amount of dilation (Ohinishi & Dharmaratne 1990, Van Sint Jan 1990). Therefore, an inevitable increase in normal stress results, the value of which depends upon the stiffness. The joint dilation is observed to be greater under a low value of initial normal stress (σ_{no}) in CNS testing, but, still smaller than under CNL condition for the same σ_{no} (Onishi & Dharmaratne 1990). Similar to the CNL condition, the asperities undergo shearing at higher σ_{no}, thereby indicating a smaller joint dilation. It is also interesting to note that the dilation response follows the shape of the asperities during shearing, if the asperities do not break (Van Sint Jan 1990).

2.5.3   *Effect of normal stiffness on joint shear behaviour*

Obert et al. (1976) showed that an increase in stiffness reduces the joint dilation, thereby increasing the normal stress with shear displacement. The peak shear stress also increases with an increase in normal stiffness, and the stress-strain behaviour is characterised by a well defined peak. Similar results are reported by Benmokrane & Ballivy (1989) and Archambault et al. (1990) for harder types of joints tested under variable normal stiffness.

2.5.4   *Effect of stiffness on shear displacement corresponding to peak shear stress*

The degradation or shearing of asperities has a significant influence on the peak shear stress. If the shearing of asperities occurs after a long shear displacement, the peak stress is generally observed at a larger shear displacement. In general, the shear displacement corresponding to the peak shear stress is greater under low normal stress (Ohnishi & Dharmaratne 1990). The increase in normal stiffness also has a significant effect on the shear displacement. As the stiffness is increased, the shear strength is also found to increase and the peak is attained at a greater shear displacement (Van Sint Jan 1990). Leichnitz (1985) reported that the shear displacement corresponding to peak shear stress under CNS is always higher than CNL.

2.5.5   *Effect of shear rate on the strength of joint*

Crawford & Curran (1981) studied the shear displacement rate on the frictional behaviour of soft and hard rock joints under CNL. Effects of normal stress ranging from 0.62 to 2.78 MPa at a shear displacement rate of 0.05-50 mm/sec are studied. Test results show that the influence of the effect of shear rate is variable, depending mainly on the rock type and the normal stress level. In general, the shear resistance decreases with increased shear displacement rate for harder rock types. Conversely, the frictional resistance increases upto a critical shear displacement rate for softer rock types, and thereafter remains constant. Curran & Leong (1983) also showed that the frictional resistance under CNL is dependent on the shear displacement rate. It is important to note that for the first time, the effect of shear rate on soft joints under CNS has been studied by the authors, and the results are discussed later in this chapter.

The scope of this book includes an overview of the CNS mechanism on soft rock joints in relation to the parameters discussed earlier. The authors have conducted a thorough laboratory testing program on clean model soft joints having various surface profiles (e.g. regular triangular asperities, tension joints and field joints) using the CNS technique. The results of this testing program will be elucidated in the subsequent sections to address some of the key questions regarding conducting CNS testing.

2.6   TESTS ON SOFT ROCK JOINTS

The behaviour of soft joints have been examined in the laboratory using the CNS apparatus designed at University of Wollongong. Triangular joints having inclinations 9.5°, 18.5° and 26.5° have been tested in addition to the tension joints and field joints. The joint specimens have been subjected to a typical normal stiffness (k_n) of 8.5 kN/mm and a shear rate of 0.50 mm/min. The test results in relation to displacement rate, normal stiffness, initial normal stress, surface roughness, and joint types are discussed below in a simplified way.

2.6.1   *Effect of shear displacement rate*

Tests have been conducted by the authors on Type II joints ($i = 18.5°$) under a shear rate of 0.35 to 1.67 mm/min. The initial normal stress (σ_{no}) and normal stiffness (k_n) were maintained at 0.56 MPa and 8.5 kN/mm, respectively, for this group of tests. The effect of shear rate on the strength and dilation characteristics of soft clean/unfilled joints is described in the subsequent sections.

Shear stress response
Initially, joints are sheared at a low shear displacement rate of 0.35 mm/min, and the changes in shear stress with horizontal displacements are recorded. The shear rate is then gradually increased. The variations of shear stress with horizontal displacement for all the tests are plotted together in Figure 2.8 for comparison. It is

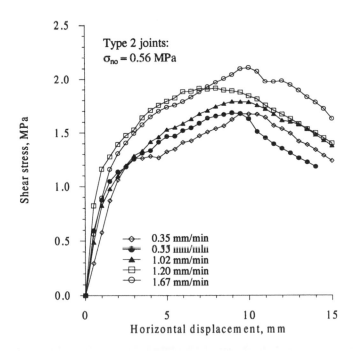

Figure 2.8. Shear stress vs horizontal displacement for Type II joints.

observed that the peak shear stress increases with the increase in shear displacement rate from 0.35 to 1.67 mm/min under the given normal stress and stiffness conditions. As discussed earlier, the frictional resistance offered by the joint surface itself becomes greater as the shear rate is increased further, and thereby, increasing the overall shear strength of joints. Similar results have been reported by Crawford & Curran (1981) for soft joints under CNL.

Variations of normal stress and dilation

The changes in normal stress and dilation with horizontal displacement are plotted for various shear displacement rates in Figure 2.9. Both the normal stress and di-

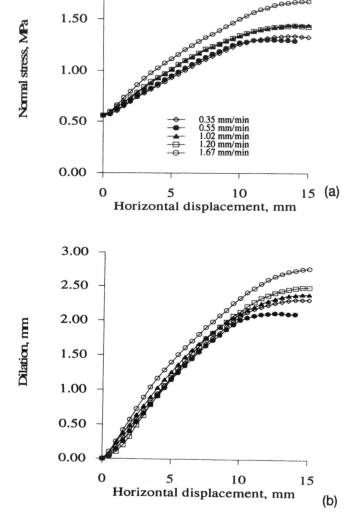

Figure 2.9. Normal stress and dilation against horizontal displacement for various shear rate of Type II joints.

lation are observed to increase with horizontal displacement, as the shear rate is increased further. The normal stress is increased by at least 25% when the shear rate is increased from 0.35 mm/min to 1.67 mm/min (Fig. 2.9a). As shown in Figure 2.9b, this increase in joint dilation is associated with a corresponding increase in normal stress.

Peak shear stress envelope under varying displacement rate

Test results indicate that the rate of shear displacement has a significant effect on the peak shear stress of soft joints. The variation of shear strength with different shear rates is plotted in Figure 2.10, and it is observed that the peak shear stress increases considerably with the increase in shear displacement rate, especially when shear rate exceeded 0.50 mm/min. Lama (1975) reported a similar behaviour for gypsum joints under constant normal load (σ_n = 2 MPa) condition. In order to define the peak shear strength envelope of soft joints more precisely, the effect of shear displacement rate on the shear behaviour should be considered more carefully, based on both CNL and CNS testing. Following these findings, all the specimens described here have been sheared at a shear (horizontal) displacement rate of 0.50 mm/min.

2.6.2 *Effect of boundary condition on shear behaviour*

Laboratory tests have been conducted by the authors on saw-tooth joints having an inclination of 9.5° and tension soft joints under CNL and CNS conditions. Initial normal stresses (σ_{no}) ranging from 0.05 to 2.43 MPa have been applied. All the specimens are sheared at a rate of 0.50 mm/min under a constant normal stiff-

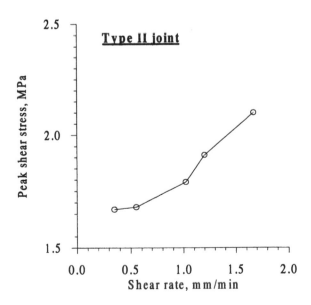

Figure 2.10. Effect of shear rate on the peak shear stress of Type II joint.

ness (k_n) of 8.5 kN/mm. Test results on saw-tooth and tension joints under CNL and CNS conditions have been recorded to compare the difference between the two types of boundary conditions, eg. free dilation boundary for CNL ($k_n = 0$) and controlled dilation boundary for CNS. The results of these tests are described in the following sections.

Effect on the shear behaviour of joints
The shear behaviour of the tension joints and triangular saw-tooth joints under CNS and CNL conditions are plotted in Figure 2.11. The left side of Figure 2.11 shows the variations in shear behaviour of natural joints, while the right hand side shows the behaviour of saw-tooth joints ($i = 9.5°$). It is observed that CNL condition ($k_n = 0$) always underestimates the peak shear stress of the joints relative to CNS, which at higher normal stresses indicates a strain softening behaviour. For both type of joints, the shear stresses are greater under CNS condition because of the increased normal stress during shearing. Similar results have also been reported by Skinas et al. (1990) and Ohinishi & Dharmaratne (1990) for harder rock joints.

Effect on horizontal displacement corresponding to peak shear stress
The change in shear stress corresponding to the initial normal stress (σ_{no}) for saw-

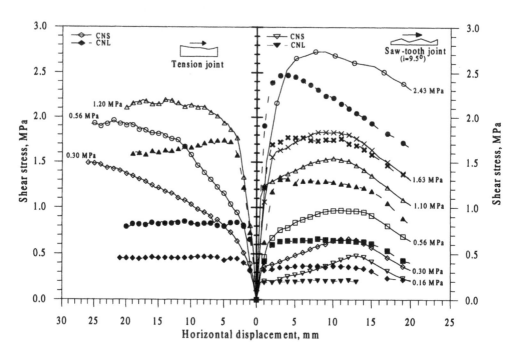

Figure 2.11. Shear behaviour of saw-tooth (Type I) and tension joints under CNL and CNS conditions.

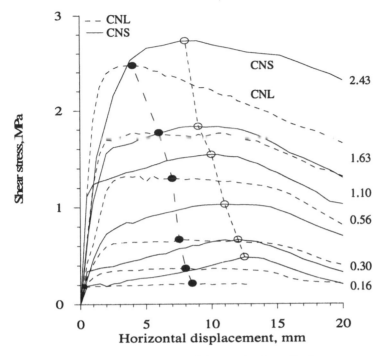

Figure 2.12. Effect of boundary condition on the shear displacement corresponding to peak shear stress for Type I joint.

tooth joints under CNL and CNS conditions are plotted in Figure 2.12. The peak shear stress is attained at a lower horizontal displacement with the increase in σ_{no}. However, the peak shear stress under CNL condition always occurs at a smaller horizontal displacement. Also, the CNL peak occurs below the CNS peak, as shown in Figure 2.12 for all the tested specimens. Similar conclusions can be drawn for tension joints.

Effect on normal stress
The variations in normal stress with horizontal displacements are plotted in Figure 2.13. The right hand side of Figure 2.13 represents the saw-tooth joints and the left hand side represents the tension joints. The normal stress increases as the asperities slide one over the other. At very low initial normal stresses (σ_{no}), the increase in normal stress is more pronounced, whereas if the σ_{no} is relatively high, the change in normal stress during shearing is gradual. The asperities shear at high normal stresses, thus the shear response is reflected by relatively flat normal stress versus horizontal displacement curves (Fig. 2.13). However, for CNL condition, it is assumed that the normal stress on the shear plane remains constant during the tests, which leads to a situation inappropriate for non-planar joints usually observed in the field.

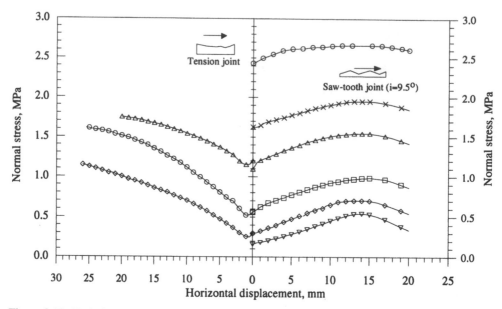

Figure 2.13. Variation in normal stresses for saw-tooth (Type1) and tension joints under CNS condition.

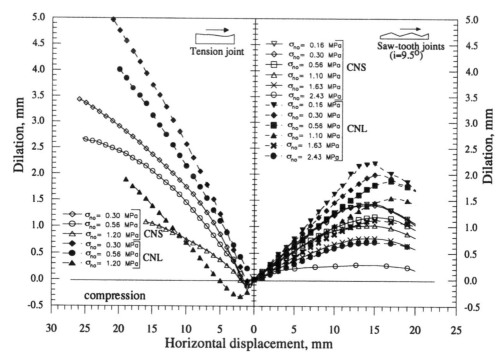

Figure 2.14. Dilation behaviour of saw-tooth (Type I) and tension joints under CNL and CNS conditions.

Effect on dilation

The variation in dilation with horizontal displacement was measured for both CNL and CNS conditions is plotted in Figure 2.14. It is observed that the CNL condition overestimates the joint dilation in comparison with the CNS condition. The dilation resulting during CNS testing is smaller, because the external springs representing the surrounding rock mass stiffness inhibits dilation to a certain extent.

Effect on strength envelopes

In order to obtain the peak shear stress envelopes for the two type of joints tested under the CNL and CNS conditions, the corresponding peak shear stress variation with normal stress for different initial normal stresses (σ_{no}) is plotted in Figure 2.15. It is evident that the CNL peak shear stress envelope for saw-tooth joints is bilinear, and it represents an upper bound for all the tests. In contrast, the CNS peak shear stress envelope can be described as linear for this particular joint type. However, the peak strength envelope for higher asperity angles tends to deviate from linearity. For the tension joints, the CNL and CNS strength envelopes are linear for the range of normal stresses investigated. If joints were tested under

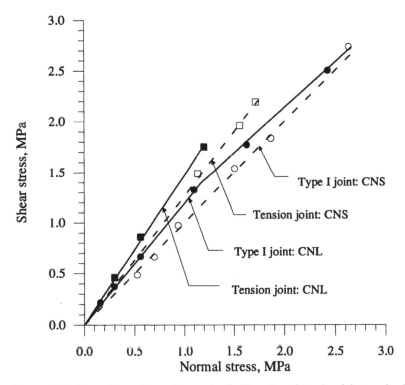

Figure 2.15. Strength envelopes for saw-tooth (Type I) and tension joints under CNL and CNS conditions.

further increase of σ_{no}, different results could be obtained. The CNL envelope for natural (tension) joints is also observed to serve as an upper bound of all the tests. Therefore, in reality, CNS test data may produce a more reliable angle of shearing resistance as required for design, especially with regard to excavations in bedded or jointed rocks.

2.6.3 *Behaviour of unfilled/clean regular joints under CNS*

Test results of two saw-tooth profiles having inclinations of 18.5° and 26.5° are presented here together with the natural (field) joints. All the joints are sheared at a displacement rate of 0.5 mm/min and subjected to normal stiffness (k_n) of 8.5 kN/mm. The initial normal stress (σ_{no}) varies from 0.05 to 2.63 MPa. The behaviour of individual joints under CNS is discussed hereafter, in relation to the shear and normal stress response, the dilation behaviour and the strength envelopes.

Shear responses of Type II ($i = 18.5°$) joints
Several CNS tests have been conducted by the authors on Type II joints under the same initial normal stress conditions as applied to Type I joints. The initial normal stress (σ_{no}) is varied from 0.05 to 2.43 MPa. As shown in Figure 2.16, a well defined peak shear stress curve is observed for all the tests, and the maximum shear stress is attained at a lower horizontal displacement as the initial normal stress is increased. The rate of increase in normal stress during shearing seems to be more pronounced under a low initial normal stress. At high initial normal stresses (e.g. σ_{no} = 2.43 MPa), significant shearing of asperities is associated with an almost constant normal stress, in comparison with the curves corresponding to lower σ_{no} values. In fact, this behaviour is similar to the conventional shearing of planar surfaces at constant normal stress.

The effect of initial normal stress on joint dilation is also investigated and is illustrated in Figure 2.16. It is obvious that dilation increases with decreasing initial normal stress. As expected, the Type II joints ($i = 18.5°$) cause a greater degree of dilation for the same normal stress levels, in comparison with the Type I ($i = 9.5°$) joints illustrated in Figure 2.16. It is important to note that the stiffness of the normal loading system, ie. the spring assembly, varies linearly once the normal stress acting on the interface becomes greater than 0.70 MPa. For $\sigma_n < 0.70$ MPa, the normal stress increment is not linear due to the normal compliance of the spring system (Fig. 2.17).

Behaviour of Type III joint ($i = 26.5°$)
Tests have also been conducted on Type III specimens under the same initial normal stresses as applied to Type I and II joints. As expected, well defined, peak shear stress curves are obtained corresponding to small shear displacements. The rate of increase in normal stress is significant at low initial normal stress, and

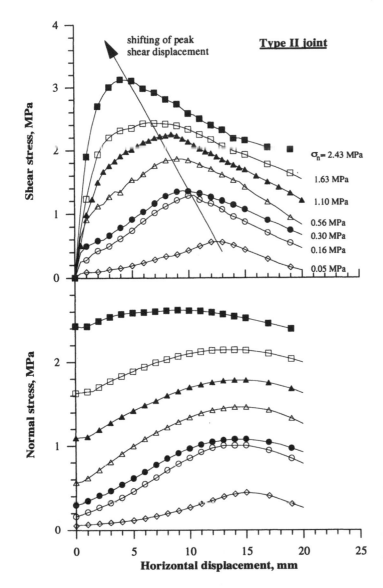

Figure 2.16. Variation of shear stress and normal stress with horizontal displacement for Type II joints ($i = 18.5°$) under CNS condition.

shearing through asperities occurred at elevated normal stress levels (Fig. 2.18). The dilation behaviour of Type III joint is similar to that of the Type I and II joints (Fig. 2.19), where a reduction in dilation is observed with the increase in initial normal stress (σ_{no}).

2.6.4 *Shear behaviour of natural (field) joints*

All natural (field) joints have been tested under an initial normal stress (σ_{no}) varying from 0.56 MPa to 2.69 MPa using a shear rate of 0.5 mm/min and exter-

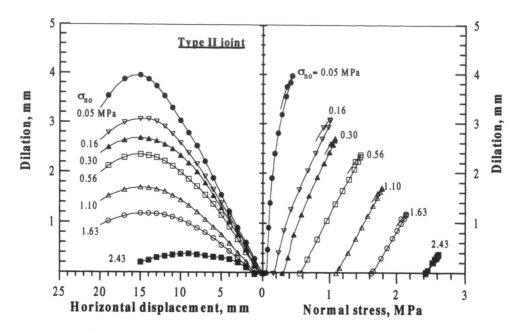

Figure 2.17. Effect of initial normal stress on dilation of Type II joints ($i = 18.5°$) under CNS condition (after Indraratna et al. 1998).

nal stiffness (k_n) of 8.5 kN/mm. The results of this series of tests are summarised below.

Shear response
The variation in shear stress with horizontal displacement is recorded at every 0.5 mm interval via a load cell, which is connected to a digital strain meter. The shear stress response for the field joints under various σ_{no} is shown in Figure 2.20 for five specimens F1 to F5. It is observed that as σ_{no} is increased, the peak shear stress is also increased. At elevated σ_{no}, the shear stress versus horizontal displacement curves show a more well defined peak. Similar behaviour is also observed for regular saw-tooth joints tested under CNS.

Normal stress
The change in normal stress with shear displacement during shearing can be monitored via a load cell connected to a digital strain meter. Figure 2.21 shows the variations in normal stress with displacement for all tests performed on the natural joints. It is observed that the normal stress increases gradually for σ_{no} of 0.56, 1.10 and 1.63 MPa due to dilation upon shearing. A decrease in normal stress results for increase in σ_{no}. This behaviour is attributed to joint compression. Similar behaviour is also observed for harder joint specimens tested under CNS condition (Ohinishi & Dharmaratne 1990).

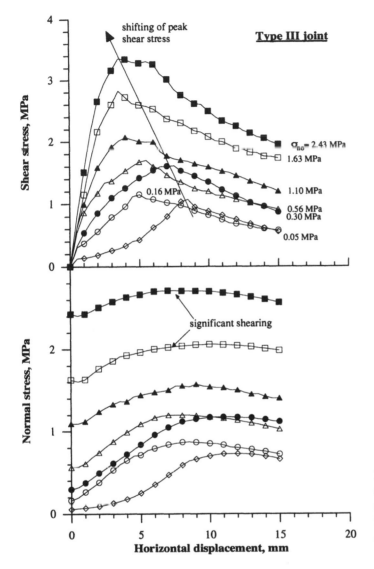

Figure 2.18. Variation of shear stress and normal stress with horizontal displacement for Type III joints ($i = 26.5°$) under CNS condition.

Dilation

The vertical movement of the joint interface can be monitored easily by a dial gauge located at the centre of the specimen top. The change in joint movement with horizontal displacements is shown in Figure 2.22 for various levels of σ_{no}. It is observed that the joint dilates under σ_{no} of 1.63 MPa, and the joint behaviour changes from dilation to compression for subsequent increase in σ_{no}. A visual observation of the joint surface indicates that the asperities shear considerably at elevated normal stresses, thereby making the shear response similar to that of the CNL testing, or sometimes indicating a shear strength smaller than CNL due to negative dilation (or compression).

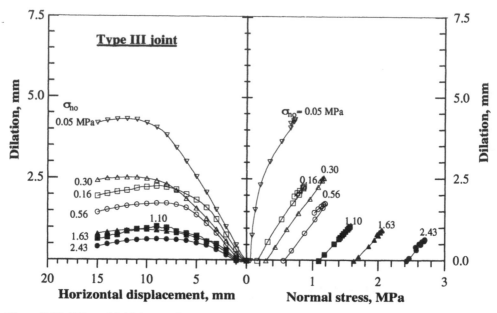

Figure 2.19. Effect of initial normal stress on dilation of Type III joints ($i = 26.5°$) under CNS condition.

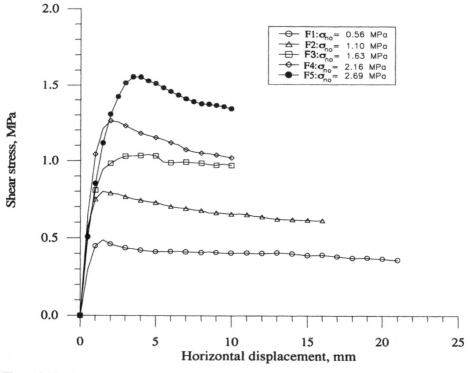

Figure 2.20. Variation of shear stress with horizontal displacement for natural (field) joints.

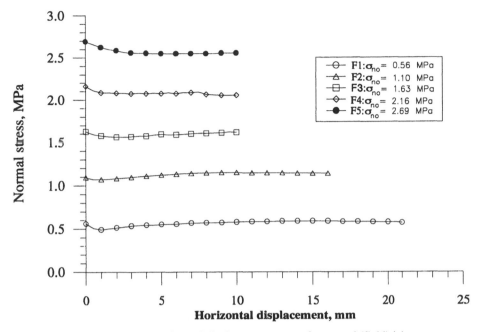

Figure 2.21. Normal stress vs horizontal displacement curves for natural (field) joints.

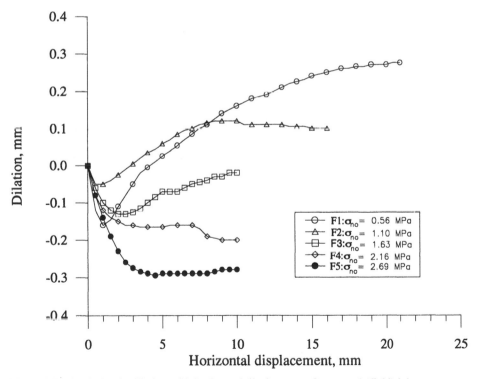

Figure 2.22. Variation in dilation with horizontal displacement for natural (field) joints.

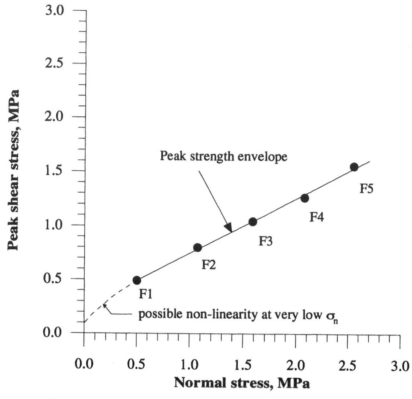

Figure 2.23. Strength envelope for natural joints (Kangaroo Valley) under CNS.

Shear strength envelope

In order to obtain the shear strength envelope, the peak shear stresses corresponding to the normal stresses are plotted in Figure 2.23 for different σ_{no} values. It is observed that a linear strength envelope is more appropriate for the sandstone joints tested under CNS for the given normal stress range, irrespective of the small variations in surface roughness.

2.6.5 *Stress-path responses of Type I, II and III joints*

The stress paths representing the change in shear stress with normal stress for different initial normal stresses for Type I, II and III joints are plotted in Figure 2.24 for comparison. It is evident from these test results that the peak shear stress increases with the increase in σ_{no}. The shear strength also increases with the increase in the asperity angle. The stress paths propagate along the strength envelope over a significant horizontal displacement at low levels of normal stress (Fig. 2.24). This implies that the shearing of asperities takes place over a greater shear displacement at lower levels of normal stress. Thus, a few tests under CNS condition are sufficient to predict the strength envelope, whereas a considerable num-

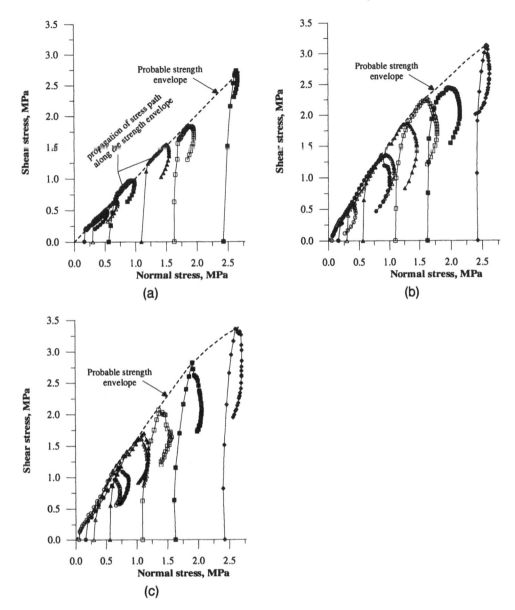

Figure 2.24. Stress-path plots for: a) Type I, b) Type II, and c) Type III joints for various σ_{no}.

ber of CNL tests are required to establish their associated strength envelope. As the normal stress is increased further ($\sigma_{no} > 2.43$ MPa), the stress paths do not seem to propagate along the strength envelope, but just reach it at peak and then the shear stress drops rapidly. It is also important to note that the stress path propagates along a shorter 'length' for Type III joints under $\sigma_{no} = 0.16$, 0.56 and 1.10 MPa in comparison to Types I and II joints. This may be attributed to the high stress concentration around the asperities of Type III joints, which cause enhanced surface degradation, thereby shearing at smaller horizontal displacements. In general, the current test results confirm that the stress-paths propagate along the strength envelope for a greater length under low normal stress, but only for a shorter length for higher σ_{no}, where surface degradation of the asperities is inevitable during shearing.

2.6.6 *Strength envelopes for Types I, II and III joints*

The shear stress and normal stress relationships for Type I, II and III profiles are plotted in Figure 2.25 for comparison. It is observed that a non-linear shear strength envelope is more applicable for interface Types II and III, whereas Type I joints show some degree of linearity, although to a lesser degree. 'Bench mark' tests (tilt test) indicated on planar interfaces at different normal stress levels where an average basic friction angle (ϕ_b) of 37.5°. The behaviour of Types II and III joints ($i = 18.5°$ and 26.5°) represented by the non-linear envelope can be explained as follows. At low normal stresses, the apparent friction angle is significantly greater than ϕ_b because of the enhanced shearing resistance offered by the angular asperities. However, at elevated stress levels, increased degradation of asperities is associated with a reduction of the apparent friction angle, which tends

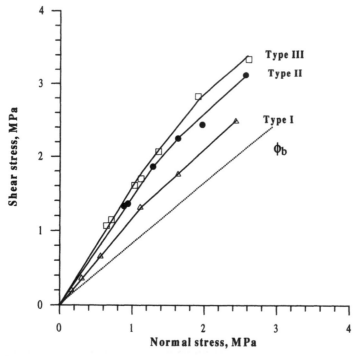

Figure 2.25. Shear strength envelopes for Types I, II and III joints under CNS condition.

to approach the basic friction angle for planar surfaces at high stress levels after considerable shearing. In contrast, Type I joints ($i = 9.5°$) are 'less frictional' due to the smaller angle of asperities, and their behaviour does not indicate such a pronounced non-linear trend. The apparent friction angle remains relatively constant at around 47°. Moreover, as discussed earlier for Type I joints in Figure 5.17a, the CNS stress paths (pre-peak) tend to follow the strength envelope in the case of Type II and III joints as well, except at high stress levels exceeding 1.5 MPa.

2.7 EMPIRICAL MODELS FOR THE PREDICTION OF SHEAR STRENGTH OF ROCK JOINTS

Patton (1966) conducted a series of tests on regular saw-teeth artificial joints under constant normal load conditions (CNL). These test results fit very well to a bilinear shear strength envelope which can be rewritten in the following forms:
– For asperity sliding:

$$\tau_{p(CNL)} = \sigma_{n(CNL)} \times \tan(\phi_b + i_0) \qquad (2.1)$$

– For asperity shearing:

$$\tau_{p(CNL)} = c + \sigma_{n(CNL)} \times \tan(\phi_b) \qquad (2.2)$$

where CNL = constant normal load condition, τ_p = peak shear stress, σ_n = normal stress, ϕ_b = basic friction angle, c = cohesion intercept, and i_0 = initial asperity angle. According to Patton (1966), the sliding of asperities takes place under low normal stress, but after a certain magnitude, shearing through asperities takes place. In contrast, other researchers considered simultaneous sliding and shearing to obtain different strength envelopes (Barton 1973, Maksimovic 1996). It has been observed that the peak shear strength predicted by Patton's model at low-medium normal stress generally overestimates the actual strength.

Barton (1973) introduced a non-linear strength envelope for non-planar rock joints for Constant Normal Load (CNL) condition as:

$$\left(\frac{\tau_p}{\sigma_n}\right)_{CNL} = \tan\left[\phi_b + JRC \times \log_{10}\left(\frac{\sigma_c}{\sigma_{n(CNL)}}\right)\right] \qquad (2.3)$$

where $\phi_b = \psi - (d_n + s_n)$, d_n = peak dilation angle which decreases with the increase in normal stress and s_n = angle due to shearing of asperities which increases with the increase of the normal stress as more surface degradation occurs. JRC = Joint Roughness Coefficient, and σ_c = uniaxial compression strength.

The method suggested by Xie & Pariseau (1992) can be used to define the value of JRC for the Types I, II and III saw-teeth profiles in the current study, as explained below:

$$JRC = 85.27 \, (D-1)^{0.57} \tag{2.4}$$

where

$$D = \frac{\log(4)}{\log\left[2\left(1+\cos\tan^{-1}\left(\dfrac{2h}{L}\right)\right)\right]}$$

In the above D = fractal dimension, h = average height of asperity, and L = average base length of asperities. Accordingly, the JRC values of 4.2, 9.0 and 13.8 have been calculated for Types I, II and III joints, respectively. These values are very close to the simplified method suggested by Macksimovic (1996), where the JRC value is considered as half of the initial asperity angle (i.e. $i_0/2$).

Assuming that at the peak shear strength under CNS condition, normal stress momentarily remains constant, Equation 2.3 (Barton 1973) can then be employed to estimate the peak shear strength. The shear strength predicted in this manner for a range of normal stresses seems to underestimate the laboratory measurements (Table 2.1). Seidel & Haberfield (1995) reported similar conclusions when Ladanyi & Archambault (1970) model was employed to predict the peak shear strength of hard concrete-soft rock joints (Table 2.2).

In order to incorporate the effect of asperities on the extent of dilation and surface degradation, Indraratna et al. (1998) suggested the following equations adopted from Jing et al. (1993) after modifying to suit the CNS condition.

$$\text{Type I:} \quad \left(\frac{i_{\tau p}}{i_0}\right) = \left[1 - \frac{\sigma_{n(CNS)}}{\sigma_c}\right]^{0.19} \tag{2.5a}$$

$$\text{Type II:} \quad \left(\frac{i_{\tau p}}{i_0}\right) = \left[1 - \frac{\sigma_{n(CNS)}}{\sigma_c}\right]^{1.5} \tag{2.5b}$$

$$\text{Type III:} \quad \left(\frac{i_{\tau p}}{i_0}\right) = \left[1 - \frac{\sigma_{n(CNS)}}{\sigma_c}\right]^{3.0} \tag{2.5c}$$

where, $i_{\tau p}$ = total dilation angle at peak shear stress under CNS condition, $\sigma_{n(CNS)}$ = normal stress corresponding to peak shear stress for a given σ_{no}, σ_c = uniaxial compression strength, i_0 = initial angle of asperity.

The increase in normal stress under CNS condition is governed by the amount of dilation of the joints during shearing. In Figure 2.26, the measured dilation (d_v)

Table 2.1. Experimental and model predicted results of peak shear stress (after Indraratna et al. 1998).

Asperity type	Initial normal stress σ_{no} (MPa)	Experimental results		Predicted peak shear stress (MPa)		
		σ_n (MPa)	τ_{peak} (MPa)	Barton (1973)	Patton (1966)	Indraratna et al. (1998)
Type I ($i = 9.5°$)						
	0.16	0.53	0.49	0.49	0.57	0.57
	0.30	0.69	0.66	0.64	0.74	0.74
	0.56	0.94	1.01	0.85	1.00	1.00
	1.10	1.50	1.54	1.32	1.61	1.60
	1.63	1.83	1.80	1.60	1.97	1.95
	2.43	2.54	2.72	2.24	2.82	2.78
Type II ($i = 18.5°$)						
	0.05	0.37	0.57	0.46	0.55	0.55
	0.16	0.85	1.30	0.95	1.26	1.18
	0.30	0.92	1.36	1.02	1.36	1.27
	0.56	1.29	1.86	1.36	1.91	1.73
	1.10	1.65	2.25	1.68	2.17	2.16
	1.63	1.97	2.44	1.96	2.41	2.52
	2.43	2.57	3.12	2.47	2.87	3.15
Type III ($i = 26.5°$)						
	0.05	0.61	1.07	0.90	1.25	1.07
	0.16	0.71	1.14	1.01	1.46	1.22
	0.30	1.05	1.61	1.37	2.15	1.69
	0.56	1.13	1.68	1.45	2.27	1.78
	1.10	1.36	2.05	1.68	2.44	2.05
	1.63	1.88	2.82	2.17	2.84	2.60
	2.43	2.58	3.35	2.79	3.38	3.20

Table 2.2. Experiemental and predicted effective friction angles for Johnston (after Seidel & Haberfield 1995).

Asperity angle, i (deg)	Measured dilation angle, δ (deg)	Basic friction angle, ϕ_b (deg)	Experimental friction angle, ϕ_{expt} (deg)	Ladanyi & Archambault (1970), ϕ (deg)
5	4	24.5	29.0	28.5
10	8.5	24.5	34.0	33.0
12.5	10.5	24.5	36.5	35.0
15	12.5	24.5	39.0	37.0
17.5	14.5	24.5	42.0	39.0
22.5	18.5	24.5	47.0	43.0
27.5	21	24.5	52.5	45.5

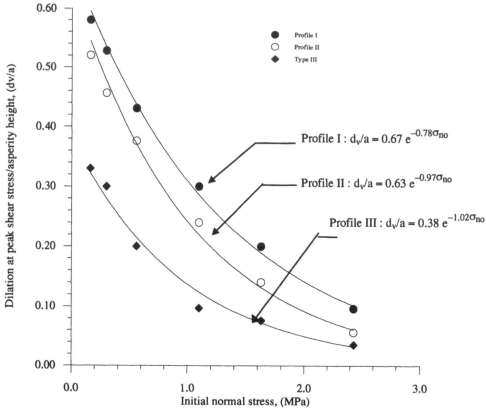

Figure 2.26. Variation of (d_v/a) with initial normal stress for profile Types I, II and III (after Indra-ratna et al. 1998; reproduced with permission from Kluwer Academic Publishers).

is divided by the height of the asperity (a) to demonstrate that the normalised ratio (d_v/a) has a unique relationship with the initial normal stress (σ_{no}) for a given joint profile. It is verified that an exponential relationship exists between the ratio, d_v/a and σ_{no}, represented by the following empirical equations:

$$\text{Joint profile Type I: } \frac{d_v}{a} = 0.67 \exp\left(-0.78\sigma_{no}\right) \tag{2.6a}$$

$$\text{Joint profile Type II: } \frac{d_v}{a} = 0.63 \exp\left(-0.97\sigma_{no}\right) \tag{2.6b}$$

$$\text{Joint profile Type III: } \frac{d_v}{a} = 0.38 \exp\left(-1.02\sigma_{no}\right) \tag{2.6c}$$

The normal stress $\sigma_{n(CNS)}$ corresponding to peak shear stress under Constant

Normal Stiffness (CNS) condition can be computed by knowing the associated dilation and normal stiffness of the joints.

Once the angle $i_{\tau p}$ is known, the total friction angle (ϕ) corresponding to the peak shear stress can be evaluated from ($\phi_b + i_{\tau p}$). The peak shear strength can then be obtained by replacing i_0 in Equation 2.1 with the value of $i_{\tau p}$ obtained from Equation 2.5. Based on this analysis, Indraratna et al. (1998) proposed the following strength envelope for CNS testing of soft joints:

$$\left(\frac{\tau_p}{\sigma_n} \right)_{CNS} = \tan \left[\phi_b + i_0 \left(1 - \frac{\sigma_{n(CNS)}}{\sigma_c} \right)^\beta \right] \tag{2.7}$$

where, $\sigma_{n(CNS)} = (\sigma_{no} + k.d_v/A)$ = normal stress corresponding to peak shear stress for a given σ_{no} under constant normal stiffness condition, k = normal stiffness (kN/mm), d_v = dilation corresponding to peak shear stress (mm), A = joint surface area (mm^2) and β = surface property which accounts the degradation of joints.

Equation 2.7 is employed to predict the peak shear strength for Types I, II and III profiles for the range of σ_{no} from 0.05 to 2.43 MPa. It is verified that the proposed model predicts the shear strength more accurately, especially in the low to medium stress range than other models (Table 2.1). The stresses obtained from Equations 2.1 and 2.2 (Patton 1966) and Equation 2.7 proposed by Indraratna et al. (1998) are plotted together with the experimental results as shown in Figure 2.27. It is evident that the proposed non-linear equation describes the peak shear strength envelope more accurately than the bi-linear Patton's model, for constant normal stiffness condition. However, this model is not capable of describing the surface roughness of more irregular joints such as field joints which require more parameters to define their surface characteristics. In order to address this issue, Indraratna et al. (1999) developed a shear strength model based on Fourier Transform method, energy balance principle and hyperbolic stress-strain relationship. Readers should follow Chapter 4 for details of this model.

2.8 SUMMARY OF BEHAVIOUR OF UNFILLED/CLEAN JOINTS

The experimental observations, analysis and empirical model predictions of the shear behaviour of saw-tooth and natural/tension joints under CNL and CNS conditions were discussed in detail. The most important aspects of the shear behaviour are summarised below.

2.8.1 *Effect of shear rate on shear behaviour of joints under CNS*

– The rate of shearing has significant effect on the shear behaviour of soft joints.

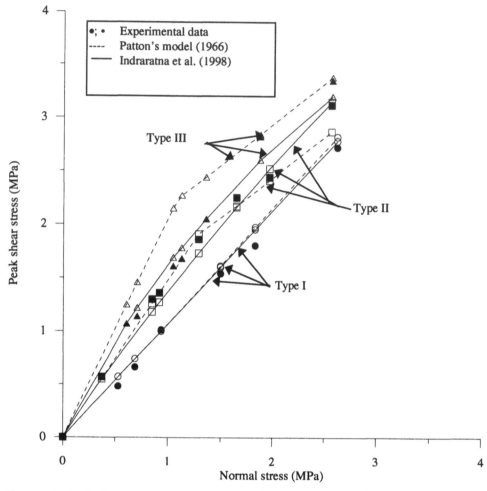

Figure 2.27. Peak shear strength envelopes for Profile Types I, II and III (after Indraratna et al. 1998; reproduced with permission from Kluwer Academic Publishers).

The shear stress and normal stress are observed to increase with the increase in shearing rate.
– A shearing rate of less than 0.5 mm/min is observed to have little or insignificant effect on the peak shear strength in comparison to a higher shearing rate.

2.8.2 *Effect of boundary condition on shear behaviour*

– CNS condition yields a higher peak shear stress than that under CNL condition for the same σ_{no}.
– Horizontal displacement corresponding to peak shear stress decreases with the increasing normal stress for CNL and CNS. However, the CNL shear displace-

ment corresponding to peak shear stress is always smaller than that of CNS.
– CNL dilation is always greater than that associated with CNS for a given initial normal stress, σ_{no}.
– CNL strength envelope always represents an upper bound for the shear stress vs normal stress data.

2.8.3 *Shear behaviour of soft unfilled joint under CNS*

– The shear strength of joints increases with the increase in asperity angle, i.e. joint roughness.
– The stress paths developed under low normal stress propagate along the strength envelope over a greater shear displacement. As the normal stress increases, the stress paths tend to reach the strength envelope briefly, but they do not propagate along the envelope. The subsequent sudden drop of the shear stress (at elevated normal stress) is associated with the shearing of asperities.
– The strength envelopes under CNS condition show a non-linear behaviour, and produce a greater apparent frictional angle ($\phi + i$) associated with the increased asperity angle (i).
– Natural joints show an increase in shear strength with the increase in normal stress. However, the natural joints are more compressive than the saw-tooth joints under a given initial normal stress. This may be due to more point contacts existing within irregular joints causing plastic deformation.
– Strength envelope of the natural joints observed in this study is linear. However, non-linear envelopes may be obtained for different degree of surface roughness and joint strength.
– Empirical methods available for predicting shear strength under CNL are not capable of estimating strength under CNS condition.
– A revised shear strength model is required to predict the dilation-strength characteristics of natural joints.

REFERENCES

Archambault, G., Fortin, M., Gill, D.E., Aubertin, M. & Ladanyi, B. 1990. Experimental investigations for an algorithm simulating the effect of variable normal stiffness on discontinuities shear strength. *Rock Joints,* Barton & Stephansson (eds), Balkema Publisher, Rotterdam, pp. 141-148.
Bandis, S., Lumsden, A.C. & Barton, N.R. 1981. Experimental studies of scale effects on the shear behaviour of rock joints. *Int. J. Rock Mech. Min. Sci. & Geomech. Abstr.,* 18: 1-21.
Barton, N. 1973. Review of a new shear strength criterion for rock joints. *Engineering geology,* 7: 287-332.
Barton, N. 1976. Rock mechanics review, the shear strength of rock and rock joints. *Int. J. Rock Mech. Min. Sci. & Geomech. Abstr.,* 13: 255-279.
Barton, N.R. 1986. Deformation phenomena in jointed rock. *Geotechnique,* 36(2): 147-167.

Barton, N. & Choubey, V. 1977. The shear strength of rock joints in theory and practice. *Rock Mechanics*, 10: 1-54.

Benmokrane, B. & Ballivy, G. 1989. Laboratory study of shear behaviour of rock joints under constant normal stiffness conditions. In *Rock Mechanics as a Guide for Efficient Utilization of Natural Resources*, Khair (ed.), Balkema Publishers, Rotterdam, pp. 899-906.

Brady, B.H.G. & Brown, E.T. 1993. *Rock mechanics for underground mining* (2nd edition). Chapman & Hall, 571 p.

Crawford, A.M. & Curran, J.H. 1981. The influence of shear velocity on the frictional resistance of rock discontinuities. *Int. J. Rock Mech. Min. Sci. & Geomech. Abstr.*, 18: 505-515.

Curran, J.H. & Leong, P.K. 1983. Influence of shear velocity on rock joint strength. *5th Int. Cong. Rock Mech., ISRM*, Melbourne, 1: 235-240.

Geological Survey of New South Wales 1974. Geology of the Wollongong, Kiama & Robertson 1:50000 Sheets: 9029-II 9028-I & IV, Department of Mines, H. Bowman (ed.), p.179.

Goodman, R.E. 1976. *Methods of geological engineering*. West Publishing company, 472 p.

Haberfield, C.M. & Seidel, J.P. 1998. Some recent advances in the modelling of soft rock joints in direct shear. *Proc. Int. Conf. Geomech./Ground Control in Mining & Underground Construction*, Wollongong, Aziz, N. & Indraratna, B. (eds), 1: 71-84.

Hencher, S.R. & Richards, L.R. 1989. Laboratory direct shear testing of rock discontinuities. *Ground Engineering*, pp. 24-31.

Hoek, E. 1977. Rock mechanics laboratory testing in the context of a consulting engineering organisation. *Int. J. Rock Mech. Min. Sci. & Geomech. Abstr.*, 14: 93-101.

Hoek, E. 1983. Strength of jointed rock masses. *Geotechnique*, 33(3): 187-223.

Hoek, E. 1990. Estimating Mohr-Couloumb friction and cohesion values from the Hoek-Brown failure criterion. *Int. J. Rock Mech. Min. Sci. & Geomech. Abstr.*, 27: 227-229.

Hoek, E. & Brown, E.T. 1980. *Underground excavation in rock*. The Institution of Mining & Metallurgy, UK.

Indraratna, B. 1987. Application of fully grouted bolts in yielding soft rocks. Ph.D Thesis, Uni. of Alberta, Canada, 308 p.

Indraratna, B. 1990. Development and applications of a synthetic material to simulate soft sedimentry rocks. *Geotechnique*, 40(2): 189-200.

Indraratna, B., Haque, A. & Aziz, N. 1998. Laboratory modelling of shear behaviour of soft joints under constant normal stiffness condition. *J. Geotechnical & Geological Engineering*, 16: 17-44.

Indraratna, B., Haque, A. & Aziz, N. 1999. Shear behaviour of idealised infilled joints under constant normal stiffness.*Geotechnique*, 49(3): 331-355.

Jing, L., Stephansson, O. & Nordlund, E. 1993. Study of rock joints under cyclic loading conditions. *Rock Mech. & Rock Engng.*, 26(3): 215-232.

Johnston, I.W. & Lam, T.S.K. 1989. Shear behaviour of regular triangular concrete/rock joints-analysis. *J. Geotech. Eng., ASCE*, 115(5): 711-727.

Johnston, I.W., Lam, T.S.K. & Williams, A.F. 1987. Constant normal stiffness direct shear testing for socketed pile design in weak rock. *Geotechnique*, 37(1): 83-89.

Kulatilake, P.H.S.W. 1992. Joint network modelling and some scale effects in rock masses. *J. of Geomechanics*, Balkema Publishers, Rotterdam, 91: 139-151.

Kulatilake, P.H.S.W., Shou, G., Huang, T.H. & Morgan, R.M. 1995. New peak shear strength criteria for anisotropic rock joints. *Int. J. Rock. Mech. Min. Sci. & Geomech. Abstr.*, 32(7): 673-697.

Ladanyi, B. & Archambault, G. 1970. Simulation of shear behaviour of a jointed rock mass. *Proc. 11th Symp. on Rock Mechanics*, Urbana, Illinois, pp. 105-125.

Lam, T.S.K. & Johnston, I.W. 1989. Shear behaviour of regular triangular concrete/rock joints-evaluation. *J. Geotech. Eng., ASCE*, 115(5): 728-740.

Lama, R.D. 1975. The concept of creep of jointed rocks and the status of research project A-6. SFB77, Jahresbericht, 1974. *Inst. Soil Mech. & Rock Mech., Uni. Karlsruhe, Karlsruhe.*

Lama, R.D. 1978. Influence of clay fillings on shear behaviour of joints. *Proc. 3rd Congr. Int. Assoc. Eng. Geol.*, Madrid 2, pp. 27-34.

Leichnitz, W. 1985. Mechanical properties of rock joints. *Int. J. Rock Mech. Min. Sci. & Geomech. Abstr.*, 22(5): 313-321.

Maksimovic, M. 1996. The shear strength components of a rough rock joint. *Int. J. Rock. Mech. Min. Sci. & Geomech. Abstr.*, 33(8): 769-783.

Obert, L., Brady, B.T. & Schmechel, F.W. 1976. The effect of normal stiffness on the shear resistence of rocks. *Rock Mech. & Rock Eng.*, 8: 57-72.

Ohnishi, Y. & Dharmaratne, P.G.R. 1990. Shear behaviour of physical models of rock joints under constant normal stiffness conditions. *Rock Joints*, Barton & Stephansson (eds), Balkema Publisher, Rotterdam, pp. 267-273.

Patton, F.D. 1966. Multiple modes of shear failure in rocks. *Proc. 1st Cong. ISRM*, Lisbon, pp. 509-513.

Pells, P.J.N. & Turner, R.M. 1978. Theoretical and model studies related to footings and piles on rock. Uni. Sydney research report R314, March.

Saeb, S. & Amadei, B. 1992. Modelling rock joints under shear and normal loading. *Int. J. Rock Mech. Min. Sci. & Geomech. Abstr.*, 29(3): 267-278.

Seidel, J.P. & Haberfield, C.M. 1995. The application of energy principles to the determination of the sliding resistence of rock joints. *Rock Mech. & Rock Eng.*, 28(4): 211-226.

Skinas, C.A., Bandis, S.C. & Demiris, C.A. 1990. Experimental investigations and modelling of rock joint behaviour under constant stiffness. *Rock Joints*, Barton & Stephansson (eds), Balkema Publisher, Rotterdam, pp. 301-307.

Van Sint Jan, M.L. 1990. Shear tests of model rock joints under stiff normal loading. *Rock Joints* (Barton & Stephansson eds), Balkema Publisher, Rotterdam, pp. 323-327.

Xie, H. & Pariseau, W.G. 1992. Fractal estimation of joint roughness coefficients. In *Proc. Int. Conf. on Fractured and Jointed Rock Masses, Lake Tahoe, California, USA*, Myer, L.R., Tsang, C.F., Cook, N.G.W. & Goodman, R.E. (eds), Balkema Publishers, Rotterdam. 3-5 June 1992, pp. 125-131.

CHAPTER 3

Infilled rock joint behaviour

3.1 INFLUENCE OF INFILL ON ROCK JOINT SHEAR STRENGTH

The presence of joints in the rock mass plays an important role on the overall shear and deformability behaviour of the rock, as well as in-situ stress and hydrogeological conditions. The jointed rock mass strength is considerably smaller than the intact rock mass strength. Therefore, for designing underground excavations and for determining rock slope stability, usage of appropriate shear strength parameters is essential. The choice of the correct shear strength parameters becomes more difficult for joints in hard rock filled with weak or loose material. The mechanical properties of the joints are to a great extent dependent upon whether the joints are clean and closed, or open and filled with infill material. The infill material may sometimes act as a cementing bond (sealant) to the joints, and in such cases, it is rarely regarded as a joint. The infill materials may consist of partially loose to completely loose cohesionless soils (e.g. sand, coarse fragmentary material etc.) which are deposited in open joints between the two surfaces of the joints. It may also be produced as a result of weathering and decomposition of the joint wall itself. Typical infill materials existing within joint interfaces can be divided into the following four categories (Lama 1978):
– Loose material brought from the surface such as sand, clay etc.,
– Deposition by ground water flow containing products of leaching of calcareous or ferruginous rocks,
– Loose material from tectonically crushed rock, and
– Products of decomposition and weathering of joints.
The thickness of the infill material may vary from a fraction of a micron to several millimeters. In tectonically crushed zones, the infill thickness may exceed several meters. The cohesion and frictional characteristics of the joints are dependent upon the properties of the infill material, the expected shear displacement, the nature of the interface surfaces (rough or smooth) and the thickness of the infill.

In the case of planar joints, the thickness of the infill material does not play any significant role on the shear behaviour, as long as the particle sizes of the infill material are sufficiently smaller than the infill thickness, so that their movement and rearrangement during shear are not constrained by the joint walls. The frictional behaviour of the joint would therefore, be that of the infill material. In the case of rough joints, the interaction between the two walls of the joint would take place depending upon the geometry of the joint surface and the thickness of the infill material. When the infill thickness is sufficiently large (i.e. more than twice the asperity height), there will be no interaction between the joint walls. Hence, the frictional behaviour of the joint will be represented by the infill alone. Sometimes, the strength of the infilled joint is considered smaller than that of the infill material as observed by Kanji (1974). However, for infill thickness smaller than twice the asperity height, the interaction between the asperities and the infill material will influence the shear behaviour.

As various infilled joints behave differently to each other, researchers have evaluated extensively, the shear strength parameters of both natural and artificial infilled rocks joints by conducting laboratory tests during the last three decades. Test results reported to date can be grouped as follows:
– Natural infilled joints tested for different surface profiles under CNL,
– Artificial rough infilled joints tested under CNL condition,
– Flat (saw cut surface) infilled joints tested under CNL, and
– Modelling of infilled joint shear behaviour.

The widely used test condition for the evaluation of infilled joint properties is the conventional direct shear test (CNL), in which the normal stress remains unchanged during shearing, i.e. zero normal stiffness ($k = 0$). It was explained in Chapter 2 that in the shear behaviour of rock-socketed piles, underground excavations in jointed rock mass and bolted joints, the normal stress would no longer remain constant during shearing. Therefore, the CNL shear behaviour investigated in the past to obtain the shear strength parameters for infilled joints may lead to unacceptable results. Thus, one of the important aim of the authors in this book is to discuss the corresponding shear strength parameters involved in modelling infilled joints under constant normal stiffness conditions. In the past, as mentioned earlier, researchers have focussed mainly on direct shear testing with constant normal load during shearing, which has provided limited knowledge on the underground shear behaviour of laminated and infilled joints. In a recent paper by de Toledo & de Freitas (1993), in which previous work on infilled joints is reviewed, they point out the importance of CNS testing on the shear behaviour of infilled joints.

3.2 FACTORS CONTROLLING INFILLED JOINT SHEAR STRENGTH

As for clean joints, the shape, size, degree of roughness and number of contacts between the surfaces control the mechanical properties. However, discontinuities existing in the rock mass may be filled as well. The infill material may be related to the origin of the fracture itself with subsequent tectonical actions (e.g. some milonites) or often, directly or indirectly related to the environmental conditions (e.g. filling material carried by water flows, by gravity or resulting from the fracturing and weathering of the rock material surface etc.). The infill material may also be a mineralisation or crystallisation of minerals (e.g. calcite, quartz, mica) or simply transported silt or clay. Soils may be sand or gravel having frictional properties, or a fine material such as clay or silt with cohesive properties. Considering such a wider range of infill material in the field, numerous research studies have been conducted in the past under CNL condition. These studies indicate that the following parameters significantly control the shear behaviour of infilled joints:

– Joint type (ie. natural or model joint),
– Type and thickness of infill material,
– Drainage condition (i.e. drained or undrained test),
– Interface boundary condition of the infilled joint,
– Infill-rock interaction, and
– Stiffness of the shear apparatus.

The effect of each of the above factors has been explained in depth in the following sections in relation to the strength and dilation behaviour of joints. Most of the previous studies have been conducted under CNL condition, which have provided a solid base to launch the authors' research under CNS condition at the University of Wollongong.

3.2.1 *Effect of joint type on shear behaviour*

In the past, laboratory tests have been conducted on natural and model joints produced precisely to match the same joint surface geometry, and without much variation in the material properties. In order to investigate the infilled joint shear behaviour, rocks of medium strength (e.g. sandstone and other sedimentary types) have been widely used by several researchers (Tulinov & Molokov 1971, Barla et al. 1985, Xu 1989), although harder rocks (e.g. basalt, granite etc.) have also been tested (Kanji 1974, Pereira 1990a, b). Concrete and casting plaster have also been popular in casting a large number of specimens in the laboratory. Surface profiles of regular and irregular shapes have been used in the past. The aim of laboratory investigation is to establish the likely shearing mechanisms involved when testing infilled joints. It is basically impossible to use natural joints for systematic laboratory testing because of the problem of obtaining the same surface geometry for repeated tests. Therefore, it is always better to match the surface profile of the natural joints using laboratory models. Idealised joints enable representation of specific joint properties, yet methods of testing of simulated joints may face serious shortcomings, if the infill material properties are not well represented.

3.2.2 *Infill type and thickness*

Besides the properties of the constituent materials, the infill thickness is perhaps the most important parameter controlling the strength of the joint. Several investi-

the most important parameter controlling the strength of the joint. Several investigations have revealed that the thicker the infill the lower the joint strength (Goodman 1970, Kanji 1974, Lama 1978, Phien-Wej et al. 1990, Papaliangas et al. 1993, de Toledo & de Freitas 1993). When the fill thickness is greater than the asperity height, this trend still seems to prevail in some cases. Previous test results (Kanji 1974) show that in some cases, the joint shear strength could be smaller than that of the infill alone. Various tests have been conducted by several researchers to investigate in detail the effect of infill thickness on the shear behaviour of rock joints (Goodman 1970, Kanji 1974, Ladanyi & Archambault 1977, Lama 1978, Phien-Wej et al. 1990, Papaliangas et al. 1993, de Toledo & de Freitas 1993). The test results can be classified according to the surface of the joints (e.g. flat, rough or saw-tooth) and the type of infill material used (e.g. cohesive or cohesionless).

Direct shear test results reported by Goodman (1970) on saw-tooth shaped joints filled with crushed mica reveal that the strength of the joint is greater than that of the infill alone, up to a t/a ratio of 1.25 (Fig. 3.1), where t = infill thickness, and a = asperity height.

Tulinov & Molokov (1971) carried out investigations on different rocks such as limestone, sandstone and marl using sand and clay layers of thickness 5 and 6

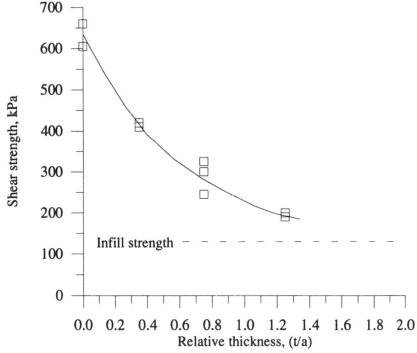

Figure 3.1. Shear strength of mica infilled joint under a normal stress of 746 kPa for various t/a ratio (after Goodman 1970).

mm. Their results show that a thin layer of sand does not have any significant influence on the frictional behaviour of harder rocks. However, in the case of softer rock (e.g. marl), its influence is significant in increasing the angle of friction. Since these tests were conducted with comparatively large infill thickness to asperity height (t/a) ratio, the interaction among the asperities probably could not take place.

Ladanyi & Archambault (1977) performed direct shear tests using kaolin clay between concrete blocks. Their results (Fig. 3.2) are similar to those of Goodman (1970). They also suggest that the steeper the asperity slope, the higher the shear strength and the greater the increase in shear strength with the decreasing t/a ratio.

Lama (1978) presented a series of laboratory tests performed on replica of tension joints filled with kaolin, in which hard gypsum was used to simulate the rock. He concluded that the strength of the joint reached that of the soil, and in some cases for t/a ratio even smaller than unity. However, for three different normal stresses shown in Figure 3.3, the envelope of the experimental data seems to approach very closely the average of the results obtained for $t/a > 1$.

Kutter & Rautenberg (1979) found that the strength of a clay filled joint increases with the surface roughness, while for a sand filled joint, a slight increase in shear strength was noticed. The overall shear resistance of the joint was reduced due to the increase of infill thickness. Wanhe et al. (1981) concluded that

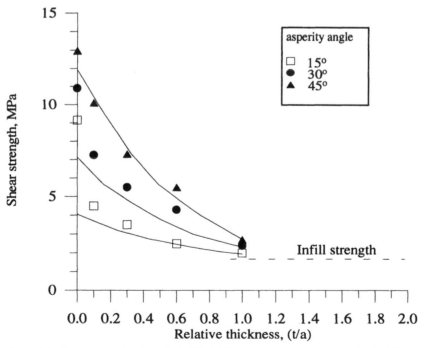

Figure 3.2. Shear strength of kaolin infilled joint under a normal stress of 8.69 MPa for different asperity angles (after Ladanyi & Archambault 1977).

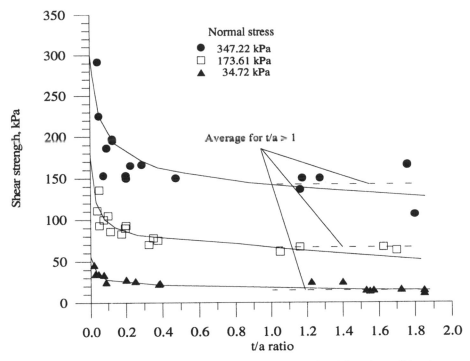

Figure 3.3. Variations in shear strength of kaolin infilled tension joints (consolidated at $\sigma_p = 350$ kPa) with increasing t/a ratio (after Lama 1978).

the shear displacement corresponding to peak shear stress gradually increased as the infill thickness was increased upto a critical value, beyond which the shear stress was controlled by the infill, and the peak shear stress was attained at a smaller shear displacement.

Phien-Wej et al. (1990) performed direct shear tests on toothed gypsum samples filled with oven dried bentonite. These test results indicate that the strength of the joint becomes equal to that of the infill when the t/a ratio approaches about 2 (Fig. 3.4).

Ehrle (1990) reported that the addition of infill material decreases the friction angle and increases the cohesion. The model rock was produced from epoxy resin mixed with a curing agent and sand. The average uniaxial compressive strength of the specimen was about 160 MPa and the joint roughness (JRC) varied from 0-10. Artificial clay consisting of sand, kaolinite, barytes, gypsum and water was used as an infill material.

Papaliangas et al. (1993) conducted a detailed testing program on plaster cement modelled joints filled with kaolin, marble dust and pulverised fuel ash (PFA). The results suggest that the strength of the joints sheared with kaolin becomes a constant at a t/a ratio of about 0.60, whereas those tested with either marble dust or fuel ash become constant at a t/a ratio between 1.25-1.50 (Fig. 3.5).

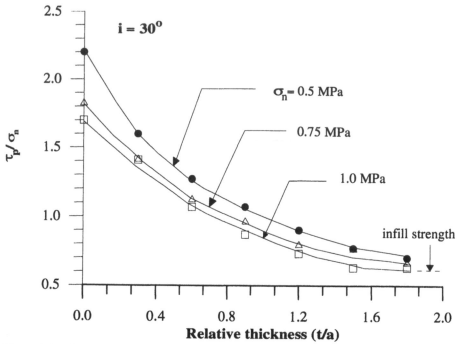

Figure 3.4. Effect of t/a ratio on peak shear stress of joints (after Phien-Wej et al. 1990).

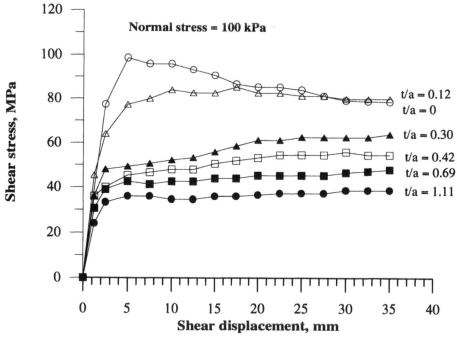

Figure 3.5. Shear stress-shear displacement graphs for different t/a ratios (replotted after Papalian-gas et al. 1993).

Test results on PFA filled modelled joints under CNL show that a sharp peak in strength occurs after a small shear displacement for a very thin layer of infill. As the *t/a* ratio is increased further, the peak (strength) becomes less well defined, and it generally occurs after a greater shear displacement. The variation of peak shear stress with *t/a* ratio is shown in Figure 3.5. A marked reduction in shear strength takes place due to the addition of even a thin layer of infill. This is because of the masking of surface texture, and perhaps also due to the introduction of rolling friction as a shear mechanism. Beyond a critical *t/a* ratio, there is a continuing but more gentle reduction in peak shear stress with *t/a* ratio due to the development of more favourable shear paths. Peak shear stress approaches a constant minimum value for *t/a* ratio between 1.25 and 1.50 (Fig. 3.6).

Tests on joints carried out with thin infill were shown to be dilatant, whereas joints with thicker infill were compressive. The dilation angles become negative (compression) for ratios of *t/a* > 0.20 to 0.30. It is also found that there is no clear dependence between the dilation angle and the normal stress level over the range of stresses employed by Papaliangas et al. (1993). The shear displacement for mobilisation of peak shear strength increases with the *t/a* ratio upto 0.75. For a *t/a* ratio between 0.75 and 1.14, a distinct strain-hardening behaviour is shown with the peak shear strength developing towards the end of the applied shear displace-

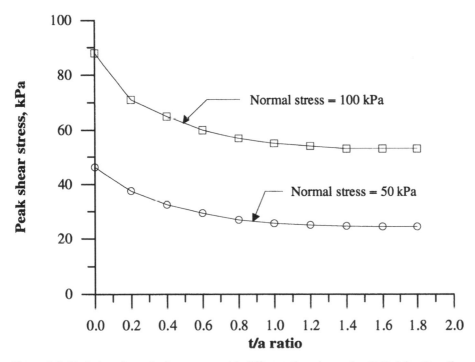

Figure 3.6. Variations in peak shear stress with different t/a ratios under CNL (after Papaliangas et al. 1993).

ment. At a t/a ratio greater than 1.14, the peak shear stress is attained at relatively small shear displacement as the infill controls the shear behaviour of joints (Fig. 3.6). In general, the shear strength decreases with increasing infill thickness. The maximum is the peak strength of the unfilled joint, and the minimum is in the range between the infill strength and the shear strength of the rock-infill interface.

de Toledo & de Freitas (1993) reported ring shear test results on toothed Penrith sandstone and Gault clay. Joints were consolidated to two different levels of stress and sheared at a constant normal stress of 1.0 MPa. Test results (Fig. 3.7) exhibit two peaks, namely the *soil peak* and the *rock peak*. The reduction in *soil peak* shear strength is observed up to t/a ratio of unity and becomes marginal beyond this t/a ratio. The *rock peak* shear strength or the ultimate strength of the joint is the same regardless of the consolidation stress of the infill, and at a t/a ratio of unity, it is greater than the strength of the soil alone. As the infill thickness tends to zero, the *rock peak* envelopes do not approach the strength envelope of the unfilled joint. It is difficult to differentiate between the two peaks when the strength differences between the infill and the rock are very small. However, it may be suggested that when the t/a ratio is greater than unity, the joint strength given by the soil may sometimes be considered equal to that of the infill alone, and not greater than it. When the shear displacement is sufficient for rock contact

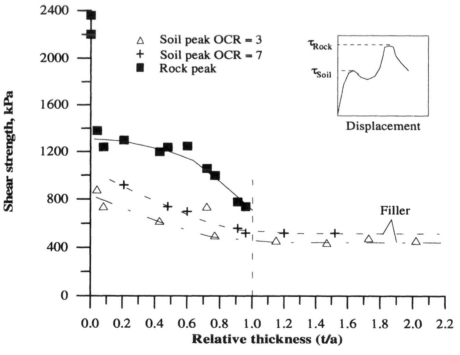

Figure 3.7. Strength of clay infilled sandstone joint tested under CNL in a ring shear device for $\sigma_n = 1.0$ MPa (after de Toledo & de Freitas 1993).

to occur during the test, the strength of the joint will also be controlled by the rock asperities.

3.2.3 *Effect of drainage condition*

The drainage condition is one of the important factors during shearing that also controls the shear behaviour of infilled joints. The drained shear strength is always greater than that of the undrained shear strength. Hence, the rate of shear displacement should be maintained according to the requirement of strength parameters, i.e. drained or undrained. Eurenius & Fagerstrom (1969) reported laboratory test results on bentonite filled chalk marl (medium to soft rock) joints under CNL. Laboratory results were also compared with in-situ test results. Shear tests were performed under consolidated undrained conditions at a strain rate of 0.6-0.7 mm/min. It was concluded from their study that the laboratory results generally agreed with those of the insitu tests.

de Toledo & de Freitas (1993) conducted tests on infilled joints at different shear rates and found that at a slower rate which simulated drained condition, the shear strength was higher in comparison with a specimen sheared at a faster (hence, undrained) shear rate.

3.2.4 *Infill boundary condition*

Another parameter that may affect the strength of the infilled joint is the condition of the infill interface as defined by the roughness of the rock wall. Kanji (1974) performed tests to investigate the effect of soil-rock interfaces on the shear strength of infilled joints. Flat saw-cut and polished surfaces of limestone and basalt were tested in shear box using different soils as the infill material. The results of the ratio of the shear strength of joint to that of the soil alone are shown in Table 3.1, which indicates that an infilled joint in some cases can be weaker than the soil that constitutes the infill material. The magnitude of the strength reduction seems to be a function of the surface roughness and of the clay mineral present in it. The drop in shear strength of a soil-rock contact occurs more sharply, and at

Table 3.1. Influence of the boundary conditions on the strength of infilled joints (after Kanji 1974).

Rock	Surface	Soil	τ_{joint}/τ_{soil}
Limestone	Saw-cut	Sandy kaolin clay	0.95
Limestone	Saw-cut	Pure kaolin	0.96
Limestone	Polished	Sandy kaolin clay	0.92
Limestone	Polished	Pure kaolin	0.88
Limestone	Polished	Illite	0.91
Limestone	Polished	Montmorillonite clay	0.76
Basalt	Polished	Montmorillonite clay	0.61

much less displacement than for the soil alone. The smoother the contact surface, the smaller is the displacement required to achieve residual strength values of the contact. This may be due to the presence of the flat, hard rock surface facilitating the orientation of clay particles along the failure plane. It was also pointed out that the rapid loss of shear strength may also lead to the possibility of premature failure, since failure may occur at smaller displacements than expected.

A joint surface with a smooth wall decreases the infilled joint strength to less than that of the infill alone. However, the infilled joint strength is higher for rougher joints as observed by Kutter & Rautenberg (1979) after conducting tests on clay filled planar to rough sandstone joints under CNL. Sun et al. (1981) performed shear box tests on concrete blocks filled with clayey sand and sandy clay with variable normal stresses and infill thickness. The failure surfaces occurred either at the top or at the bottom rock contact, or as a combination of both surfaces.

Pereira (1990b), using mainly sand filling between two flat granite blocks, also reported failure along the solid boundaries due to the rolling of sand grains. Solid boundaries affect the strength of a joint in two ways. In clay fills, sliding occurs along the contact due to particle alignment, whereas in sands, the rolling of grains seems to be the major factor responsible for weakening the joint. The friction behaviour of joints filled with coarse material is considered to be influenced by the interface between the infill material and the rock block surface. Whenever the thickness is equivalent to the average grain dimension, a situation of rolling friction is deemed to have occurred. The magnitude of the influence of surface roughness depends on the particle size of the soil. In simple form, when sand as infill material is considered, the influence of the rock boundary may become marked when its surface is smoother than the roughness of the sand surface (defined by particle size distribution) and when the dilation is reduced. As illustrated in Figure 3.8, two joints having different roughness are filled with the same sand. The joint shown in Figure 3.8a is rough enough to prevent movement of the sand-rock contact, and the sliding friction of the sand has to be overcome for failure to occur. In contrast, the joint in Figure 3.8b is smooth and allows grain rotation on the boundary, hence, only rolling friction has to be overcome for failure to occur. For the clay fill, very small roughness is liable to reduce the shear strength of joints (Kanji 1974).

3.2.5 *Infill-rock interaction*

The stiffness of the fill under shear can, in some cases, affect the strength of the joint because of the problems dealing with two materials with different mechanical properties. The role of the infill is not only to weaken the joint but also to impose a failure mechanism different to that of the unfilled joint or of the infill alone. In practice, it is assumed that the minimum shear strength of infilled rock joint is the shear strength of the infill itself. It is evident that the shear surface can

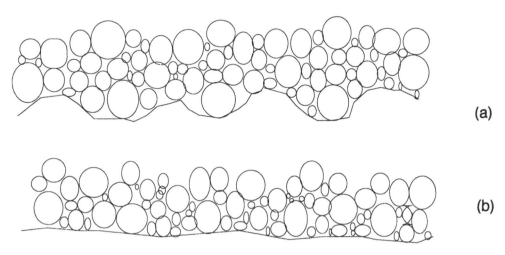

(a)

(b)

Figure 3.8. Rock joint-sand fill contact: a) Rough surface with no influence in the joint strength, and b) Smooth surface with weakening of the joint (after de Toledo & de Freitas 1993).

occur along the boundary between two different materials (Giuseppe 1970, Clark & Meyerhof 1972).

As mentioned by several researchers (Barton & Choubey 1977, de Toledo & de Freitas 1993), the shear failure of an infilled joint takes place into two stages:
– *First stage.* Shear stress and shear displacements are controlled only by the strength of the infill.
– *Second stage.* After some displacements have taken place, the two rock surfaces come into contact and the strength of the joint is governed mainly by the shape of the asperities and the strength of the rock.

Depending upon the level of the normal stress, dilation caused by the sliding of one block over the other may occur, subsequently followed by the breakage of asperities, as normally occurs in unfilled joints. When the infill thickness is greater than the asperity height, the joint can fail along a continuous surface not intercepted by the rock asperities. Thus, it could be expected that the shear strength of the joint would be equal to that of the infill alone. Therefore, it is clear that as the infill thickness becomes smaller than the asperity height, the expected shear strength of the joint tends to increase more than that of the infill alone, due to the significant influence of asperities during shearing. Results reported by Kutter & Rautenberg (1979) on clay filled sandstone joints also support the contribution of both infill and asperities for lower infill thickness.

Phien-Wej et al. (1990) reported that the asperities came into contact when testing bentonite filled joints under CNL condition, and that the infill thickness was thinner than 60-80% of the asperity height (Fig. 3.9). The shear process was accompanied by both shearing through asperities and dilation, for the given normal stress range. However, the contribution of asperities to the shear strength

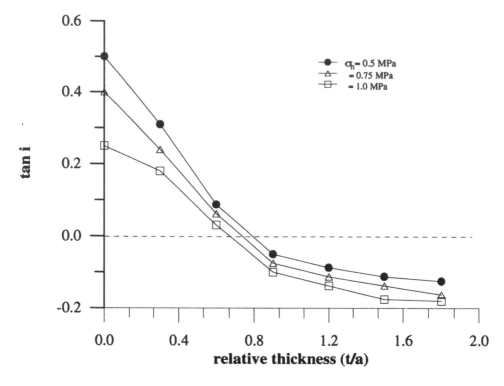

Figure 3.9. Effect of t/a ratio on dilation of joint under CNL (reproduced from Phien-Wej et al. 1990).

ceased when the infill thickness exceeded or approached the asperity height. The rougher the joint surface, the greater was the dilation, and the transition from dilation to compression occured faster. The strength of the joint attained that of the infill when t/a ratio reached 2, regardless of the applied normal stress (Fig. 3.9). Goodman (1970) studied the plaster-celite cast toothed joints filled with crushed mica tested under CNL condition. He found that the thickness of the infill needed to be at least 50% greater than the asperity height so that the joint strength would be as small as the strength of the infill itself.

The difference in stiffness between the rock and the infill facilitates progressive failure (de Freitas & de Toledo 1993). Figure 3.10 shows the distribution of the potential displacements generated by a horizontal point force applied to the infill along vertical sections and passing through the tips of the asperities, while the rock samples remain fixed. The infill of regular rock joints cannot deform uniformly in the direction of shear, and the lack of uniformity is most pronounced close to the tips of the asperities, where stress concentration occurs and facilitates the development of the failure surface. Thus, the failure surface tends to start close to the tips and edges of roughness and to propagate to the rest of the fill, presenting a shear strength smaller than that of the soil alone.

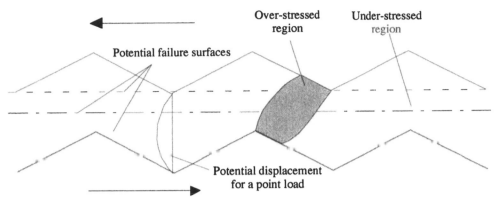

Figure 3.10. Stress concentration on a filled joint (after de Toledo & de Freitas 1993).

Pereira (1997) performed tests on sand filled Penrith sandstone specimens using a rotary shear machine. The mechanical properties of the joint material were: Joint Wall Compressive Strength (JCS) of 40-50 MPa and basic friction angle (ϕ_b) of 29-32°. Sand of river origin containing grain sizes from 1.18 mm to 0.4 mm was used as an infill material. Test results suggested that rolling friction controlled the shear mechanism, and this may be more relevant than sliding friction for discontinuities filled with dry and uniform sands, even at small displacements. However, for larger shear displacement, sliding friction becomes more pronounced. Shear strength of the filled joints was always smaller than that of unfilled joints.

3.2.6 *Effect of external stiffness*

All the test results on infilled joints reported previously were obtained by conducting conventional direct shear or ring shear tests, where shearing takes place under CNL. As pointed out in Chapter 2, this condition is particularly suitable for joints having planar surfaces, where shearing is associated with negligible dilation. In contrast, shearing of non-planar joints takes place under constant normal stiffness (CNS) condition. de Toledo & de Freitas (1993) mentioned that results based on the direct shear apparatus are relevant to conditions of constant zero normal stiffness; inevitably, different results would be obtained if the normal stiffness of the apparatus were changed. Cheng et al. (1996) studied Johnstone paste (mixture of mudstone powder, cement and water) infilled concrete/rock interfaces under CNS condition. Test results showed that unlike in CNL conditions, a very thin smear zone (thickness < 1 mm) was unable to reduce significantly the strength of the joint. Almost the same strength was observed for dry cast joints having no infill and infill thickness of 2 mm. This conclusion contradicts the findings of CNL testing reported by several researchers. However, for a thicker infill thickness, both CNS and CNL findings are similar. Shear stress vs normal stress plot of concrete/rock infilled joints revealed that the shear response is purely frictional and is independent of smear thickness.

Considering limited research on infilled joints tested under CNS, the authors conducted an extensive laboratory testing program on idealised soft toothed joints filled with commercial bentonite of various thickness, and tested under varying initial normal stress (σ_{no}). Details of these test results are discussed in this chapter to clearly present the parameters and mechanisms involved in infilled joint behaviour.

3.2.7 *Normal stress and lateral confinement*

The range of normal stresses applied during shear testing in past studies have varied from 0.02 to 15 MPa. Such a normal stress range covers a wide range of application of shear strength parameters (e.g. slope stability, design of underground excavations etc.).

When a normal stress is applied on the joint prior to testing, there is a risk of infill material emanating from the joint due to the consolidation process. In order to assess the infill joint shear behaviour properly, it is necessary to avoid side friction due to lateral support and the potential loss of infill at the same time. To solve this problem, lateral confinement may be used during the consolidation process and removed just before the shearing phase (de Toledo & de Freitas 1993). Different approaches have been reported ranging from no confinement (Pereira 1990b) to a fixed confinement (Xu 1989). Rubber membrane has also been used as a lateral confinement (Barla et al. 1985) as well as a Teflon ring (Kutter & Rautenberg 1979) or providing an independent sliding ring that maintains a closed gap while keeping friction very low (de Toledo & de Freitas 1993). The 'squeezing out' of the soft infill layer is a phenomenon only typical for laboratory conditions, but rarely occurs in the field, because of the size of joints. Therefore, it is important to select the volume of infill so that it is realistic in relation to the available shearing area in the laboratory, which is clearly limited in comparison to field conditions.

3.3   LABORATORY TESTING OF INFILLED ROCK JOINTS

The behaviour of infilled joints under CNL condition has been investigated in the past, in depth. To the authors' knowledge, one paper (Cheng et al. 1996) has dealt with the shear behaviour of infilled concrete/rock interface under CNS condition based on the initial work conducted by Cheng (1996). In order to further understand the shear behaviour of soft rock/rock infilled joints, tests have been conducted on the saw-tooth joints of inclinations 9.5° (Type I) and 18.5° (Type II) under a constant normal stiffness (k_n) of 8.5 kN/mm (or 453 kPa/mm for a joint surface area of 187.5 cm^2). Seven different series of tests (Table 3.2) have been conducted under an initial normal stress (σ_{no}) of 0.16, 0.30, 0.56 and 1.10 MPa for varying t/a ratios. All infilled joints have been sheared horizontally under drained condition at a constant rate of 0.50 mm/min. These results have also been discussed by Indraratna et al. (1999).

Table 3.2. Test condition and test series of infilled joint under CNS.

Name of series	Type of joint	σ_{no}, MPa	Infill thickness (t), mm	t/a ratio
Series I	Type I ($i = 9.5°$)	0.16	0.0	0.0
		0.16	1.0	0.4
		0.16	1.5	0.6
		0.16	2.0	0.8
		0.16	4.0	1.6
Series II	Type I	0.30	0.0	0.0
		0.30	1.5	0.6
		0.30	2.5	1.0
		0.30	3.5	1.4
		0.30	4.0	1.6
Series III	Type I	0.56	0.0	0.0
		0.56	1.5	0.6
		0.56	2.5	1.0
		0.56	3.5	1.4
		0.56	4.5	1.8
Series IV	Type I	1.10	0.0	0.0
		1.10	1.5	0.6
		1.10	2.5	1.0
		1.10	3.5	1.4
		1.10	4.5	1.8
Series V	Type II ($i = 18.5°$)	0.30	0.0	0.0
		0.30	1.5	0.3
		0.30	3.0	0.6
		0.30	5.0	1.0
		0.30	7.0	1.4
		0.30	9.0	1.8
Series VI	Type II	0.56	0.0	0.0
		0.56	1.5	0.3
		0.56	3.0	0.6
		0.56	5.0	1.0
		0.56	7.0	1.4
		0.56	9.0	1.8
Series VII	Type II	1.10	0.0	0.0
		1.10	1.5	0.3
		1.10	3.0	0.6
		1.10	5.0	1.0
		1.10	7.0	1.4
		1.10	9.0	1.8

Note: at least two tests for each t/a ratio were conducted to obtain the average mean response.

3.3.1　*Selection of infill material*

Commercial bentonite has been selected by the authors' as the appropriate infill material for the gypsum plaster based soft joints. A placement moisture content of 12 (±1)% has been ensured for all tests by keeping the specimens inside a sealed container. Direct shear tests have been conducted on the infill material for a wide range of normal stresses ranging from 0.07 to 0.45 MPa (Indraratna et al. 1999). All specimens have been sheared under drained condition at a constant rate of 0.50 mm/min., which is similar to the shear rate used for infilled joints under CNS condition. Based on these results, the shear behaviour of infill is representative of a linear Mohr-Coulomb material with a peak friction angle (ϕ_p) of 35.5° (Fig. 3.11). Phien-wej et al. (1990) have verified that bentonite is appropriate to simulate many prototype infill materials. Moreover, a friction angle of 36° has been measured by Suorineni & Tsidzi (1990) for a typical infill material in the shear zones of the Birimian system in Ghana.

3.3.2　*Preparation of infilled joint surface*

This section describes the preparation of infilled joints for laboratory testing, as

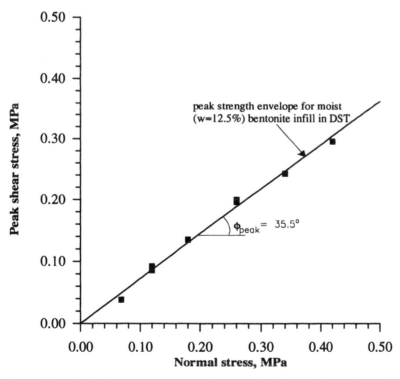

Figure 3.11. Peak shear stress envelope for bentonite infill in direct shear (after Indraratna et al. 1999).

recommended by the authors. Saw-tooth joint profile is selected, as shown in Figure 3.12. The bottom half of the saw-tooth specimen is placed inside the shear apparatus, such that the surface profile is projected slightly above the bottom half of the mould (Fig. 3.12a). An adjustable collar having the same surface profile is then attached to the top of the specimen, hence, making an enclosure over the specimen. The collar is now fixed tightly to provide the required infill thickness by precisely measuring the height of the mould at the four corner points (Fig. 3.12a). The bentonite is then placed inside the collar in small quantities and spread over the joint surface by a spatula. Once the collar is filled up, the bentonite (infill) surface can be compacted and trimmed by a flat steel plate having the same saw tooth shape. Subsequently, the collar is dismantled and the bottom specimen is placed within the shear apparatus and fixed firmly by tightening all the screws (Fig. 3.12b). The top shear box containing the upper half of the specimen is then placed over the bottom specimen, and simultaneously, the smooth lateral support plates made from stainless steel are assembled around the infill joint to prevent loss of infill during shearing (Fig. 3.12c). The change in moisture content of infill material during specimen preparation and shearing is neglected in this study. A close up view of the prepared laboratory specimen is shown in Figure 3.13. In this study, all joints were consolidated and sheared under the predetermined initial normal stress (i.e. σ_{no} = 0.16, 0.30, 0.56 and 1.10 MPa). Preshear consolidation under the applied initial normal stress (σ_{no}) would usually took between 45 minutes and 1 hour.

3.3.3 *Setting-up the specimen in the shear boxes*

The cured specimens, once they attained the room temperature, could be placed inside the top and bottom shear boxes and secured tightly by adjusting the screws. For the infilled joints, the infill surface was carefully prepared according to the previously outlined method for the bottom half of the specimen. The bottom box was then placed inside the lower specimen holder. Then the top shear box was inserted inside the upper specimen holder and screwed tightly with the top plate. A close up view of the infilled joint placed inside the shear box is shown in Figure 3.14. In order to obtain an exactly mated joint condition, the top half of the specimen was placed on the bottom half by forwarding or reversing the bottom shear box. Finally, a set of springs representing the surrounding rock mass stiffness (8.5 kN/mm) was placed above the specimen. These elastic springs maintain the Constant Normal Stiffness (CNS) condition upon loading.

3.3.4 *Application of normal load*

Before shearing the specimen, the predetermined normal load could be applied through the hydraulic jack using a manual or electric pump. The digital strain meter fitted to the normal load cell was utilised to indicate the current normal

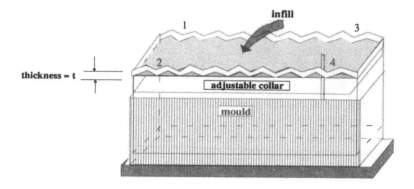

(a)   Stage I: Fixing the adjustable collar above the bottom specimen and pouring infill.

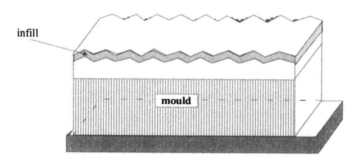

(b)   Stage II: Dismantling of collar and fixing infilled joint in position.

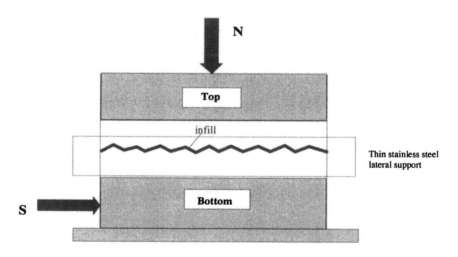

(c)   Stage III: Placement of top and bottom specimen together with lateral support.

Figure 3.12. Preparation of infill joint surface (Haque 1999).

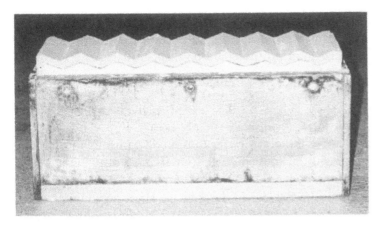

Figure 3.13. A close view of the prepared infilled joint surface.

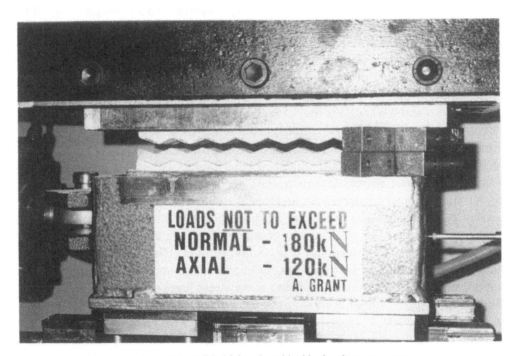

Figure 3.14. A close up view of the infilled joint placed inside the shear apparatus.

load. The vertical dial gauge fitted on the top of specimen indicated a stable reading once the specimen was lightly consolidated under an initial σ_{no}. For infilled joints, the settlement of the infill reached a constant value after 45 minutes to 1 hour from the time of application of the normal load. The normal load was then adjusted to its previous level (σ_{no}) by raising the pump pressure. Once the consolidation settlement is over, the specimen is ready to shear. At the end of the each test, the specimen was brought back to its initial position by reversing the shear direction. The shear boxes were dismantled afterwards, and the mode of failure was observed. The final joint profile could be mapped using the Coordinate Measuring Machine (CMM).

3.3.5   *Shear behaviour of Type I joints*

Tests have been conducted under the given σ_{no} values on Type I joints (asperity height = 2.5 mm) having an infill thickness ranging from 0 to 4.5 mm, which correspond to a infill thickness/asperity height (t/a) ratio from 0 to 1.80 (Haque 1999, Indraratna et al. 1999). Sections 3.3.5 to 3.3.11 describe the shear behaviour of infilled joints as also discussed by Indraratna et al. (1999). The variations in shear and normal stresses were continuously recorded with the horizontal displacements up to 16 mm, which is greater than the half base length of the triangular asperities. Test results show that the peak shear stress drops significantly in comparison with the unfilled joints due to the addition of a thin infill layer of 1.5 mm in thickness. The maximum shear stress continues to drop with the increasing infill thickness, as shown in the upper part of Figures 3.15 and 3.16. As the infill thickness (t) becomes the same as that of the asperity height (i.e. $t/a = 1.0$), the shear stress and normal stress responses remain relatively unchanged even at large displacements, for $\sigma_{no} = 0.16$ and 0.30 MPa (Figs 3.15a, b, and 3.16a, d). This indicates that the effect (contact) of asperities is reduced, and that the shear behaviour is now governed mainly by the infill. Nevertheless, the slight dilation noted on Figure 3.16g indicates that the asperities still influence behaviour at large displacements. As the infill thickness is increased further, the shear stress attains the peak quickly at very small horizontal displacement, and then continues to drop gradually.

At an infill thickness (t) exceeding 2.5 mm, the shear stress and normal stress curves represent the typical behaviour of soil infill, where the asperities no more influence the stress-strain response (Figs 3.15a, b, and 3.16a, d). The negative dilation for $t = 3.5$ mm and 4.0 mm (Figs 3.15c and 3.16g) verifies the compression or squeezing out of infill during shearing at the stress level of $\sigma_{no} = 0.16$ and 0.30 MPa. Even at higher initial normal stresses (i.e. $\sigma_{no} = 0.56$ MPa and 1.10 MPa), the shear behaviour is governed mainly by infill for $t > 2.5$ mm, as no dilation is observed during shearing (Figs 3.16h and i). In other words, it may be concluded that the change from dilatant to compressive behaviour of joints takes place as the critical infill thickness to the asperity height ratio is exceeded.

A close visual examination of the surface of sheared joints revealed that the asperities were subject to moderate to low grade damage for an infill thickness up to 2.5 mm, whereas no sign of asperity damage was observed for greater infill thickness (i.e. $t = 3.5$ mm and 4.5 mm). This implies that the t/a ratio of 1.40 (corresponding to $t = 3.5$ mm) may be considered as the 'critical ratio', beyond which the asperities (surface profile) have no effect on the shear behaviour of joints.

Based on CNL tests, de Toledo & de Freitas (1993) observed two peaks on the

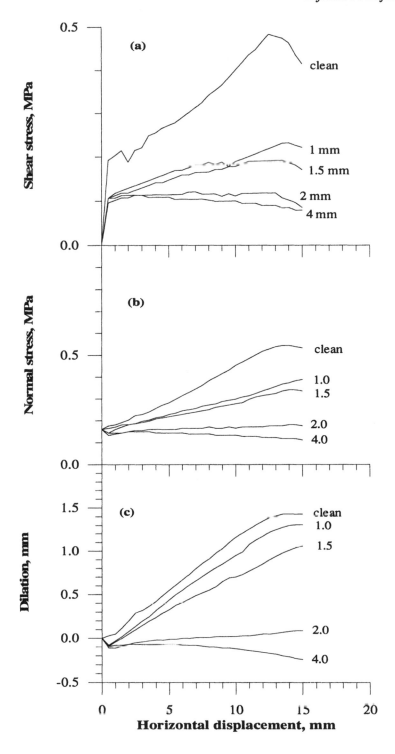

Figure 3.15. Shear behaviour of Type I infilled joint under σ_{no} = 0.16 MPa (Haque 1999).

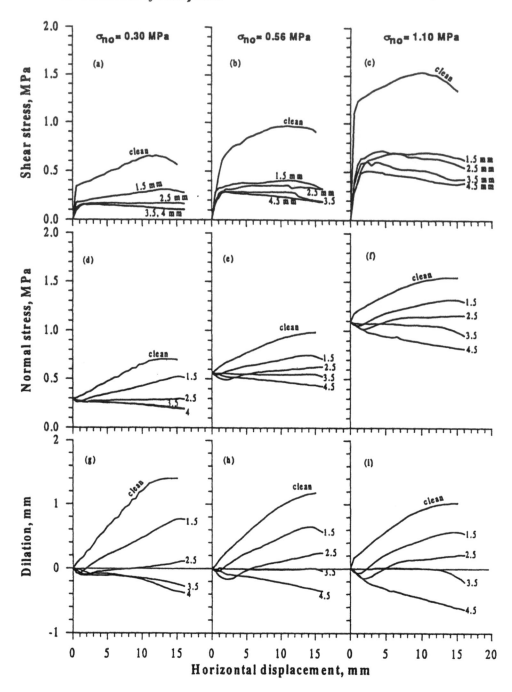

Figure 3.16. Shear behaviour of Type I infilled joint under σ_{no} = 0.30-1.10 MPa (Indraratna et al. 1999).

stress-strain response, i.e. one peak due to the infill (soil peak) and the subsequent peak due to the natural rock asperities (rock peak). However, in the CNS tests conducted in this study, only one distinct peak was observed for all tests, and this peak was associated with the asperities or infill, depending on the t/a ratio. Moreover, as the simulated joints used in this study were made of gypsum plaster, the difference between the infill (soil) peak and the 'soft' rock peak was expected to be small.

3.3.6 *Shear behaviour of Type II joints*

Tests have been conducted on Type II interface (asperity height = 5 mm) for infill thickness ranging from 0 to 9 mm under an initial normal stress (σ_{no}) of 0.30 to 1.10 MPa. The shear behaviour of these tests is plotted in Figure 3.17 along with the clean joint. The drop in peak shear stress of the infilled joints becomes insignificant as the infill thickness is increased beyond 7 mm, or as t/a exceeds 1.4 at $\sigma_{no} = 0.30$ MPa (Fig. 3.17a). For $\sigma_{no} = 0.30$ and 0.56 MPa, an increase in normal stress is observed until such time as the infill thickness exceeds 5 mm, beyond which a decrease in normal stress is noted (Figs 3.17d and 3.17e). This is associated with joint compressive behaviour (Figs 3.17g and h) suggesting that the influence of asperities is now negligible. At a greater σ_{no} value of 1.10 MPa, the effect of asperities is insignificant at $t = 9.0$ mm ($t/a = 1.8$).

The profiles of the shear planes for all the tests conducted under an initial normal stress of 0.56 MPa were estimated from the measured dilations corresponding to the horizontal displacements, and are plotted in Figure 3.18. It is important to note that this shear plane (dashed line in Fig. 3.18) is a theoretical line representing the surface on which the average sliding process takes place. It clearly shows that the shear plane passes through the asperity and infill for an infill thickness of 1.5-3 mm (or $t/a = 0.3$ to 0.6), and touches only the 'crown' of the saw toothed asperity for an infill thickness of 5 mm (or $t/a = 1.0$). For $t = 7$ mm and 9 mm, the shear plane passes through the infill material only, and the shear behaviour of such joints is fully governed by the infill alone. Therefore, a t/a ratio between 1.0 and 1.4 can be considered to be 'critical' under the given test circumstances. However, this critical ratio could increase up to 1.8 for higher σ_{no} values, exceeding 1.10 MPa.

3.3.7 *Effect of infill thickness on horizontal displacement corresponding to peak shear stress*

The shear displacements corresponding to the peak shear stress (δ_p) against various t/a ratios for Type I and II profiles are plotted in Figure 3.19. It shows very clearly that a sudden drop of δ_p takes place as the critical ratio is approached or exceeded. Similar type of behaviour was also reported by Phien-wej et al. (1990) for tests carried out under constant normal load (CNL) condition. However, for

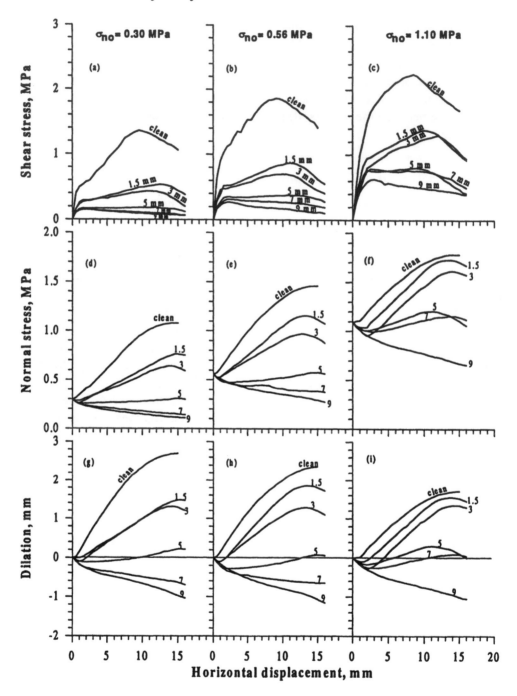

Figure 3.17. Shear behaviour of Type II Infilled joints under σ_{no} = 0.3-1.10 MPa (Indraratna et al. 1999).

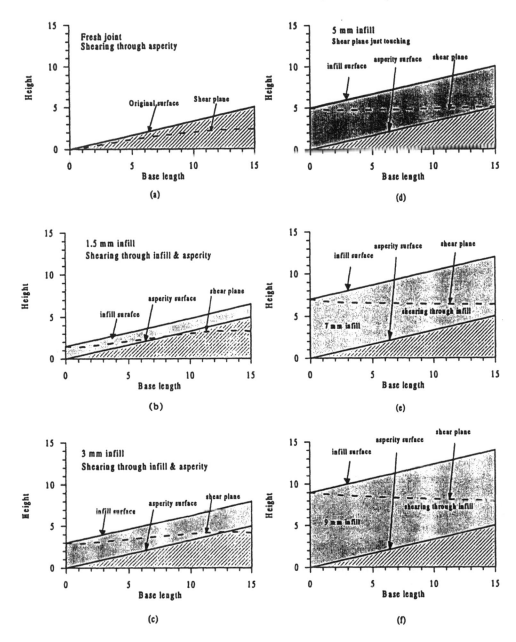

Figure 3.18. Half of the asperity surface and shear plane for infilled Type II joint sheared under $\sigma_{no} = 0.56$ MPa (Indraratna et al 1999).

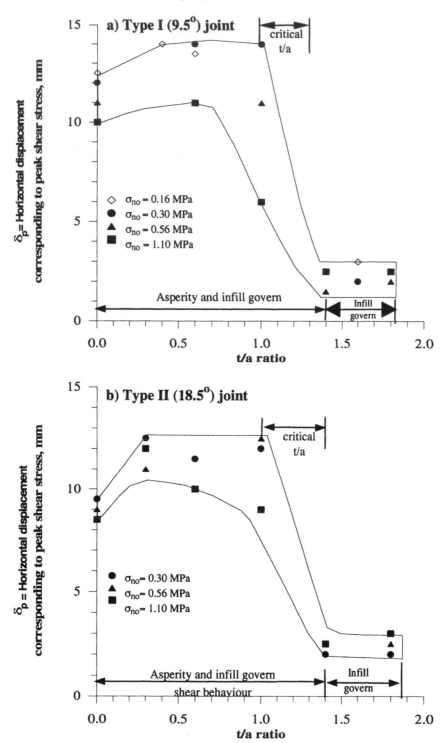

Figure 3.19. Horizontal displacement corresponding to peak shear stress against (*t/a*) ratio for various initial normal stresses (σ_{no}) for: a) Type I joint, and b) Type II joint (Indraratna et al. 1999).

CNL tests, the actual *t/a* ratios were observed to be much higher (exceeding 2.0), even at small normal stresses.

3.3.8 *Effect of infill thickness on stress-path behaviour*

The variation in shear stress with the normal stress for Type I and II profiles for various σ_{no} values is plotted in Figures 3.20 and 3.21, in order to obtain the stress-path plots. Both Type I and Type II joints indicate that once the critical *t/a* ratio is exceeded for a given value of initial normal stress (σ_{no}), the corresponding stress-path plot indicates a reduction in normal stress throughout and an early increase in

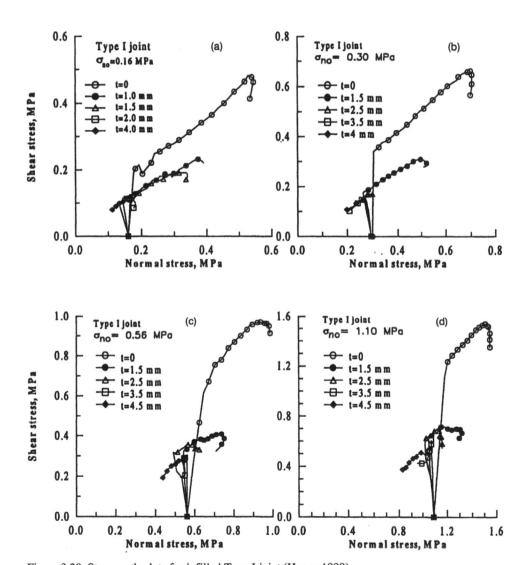

Figure 3.20. Stress path plots for infilled Type I joint (Haque 1999).

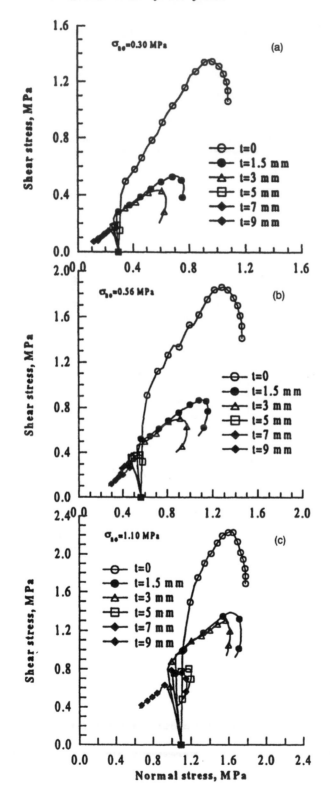

Figure 3.21. Stress-path plots for Type II infilled joints (Haque 1999).

shear stress followed by a decrement. If the critical t/a ratio is not exceeded, then the stress paths plot reveals an increase in normal stress until shearing of asperities takes place and an increase in shear stress occurs for a considerable displacement. Also, as the infill thickness is increased, the 'band width' (i.e. extent of variations of shear and normal stress) of the stress paths tends to decrease, which is representative of reduced dilation or increased compression of the overall joint. This is clearly because of the diminished contact between the joint asperities as the t/a ratio is increased.

3.3.9 *Effect of infill thickness on peak shear stress*

The peak shear stress obtained for infilled Type I and II joints is plotted against the t/a ratio for various values of σ_{no} (0.16-1.10 MPa), together with the clean joints in Figure 3.22. It is observed that the addition of little infill (say, 1.5 mm in thickness) decreases the joint strength by almost 50%. As the infill thickness is increased further, the peak shear stress continues to decrease gradually, and after a certain value of t/a ratio is reached, further decrease in strength becomes marginal. Figure 3.22 illustrates clearly that as the t/a ratio increases, the overall joint strength approaches that of the pure bentonite infill (or becomes asymptotic) particularly in the case of Type II joints, where the actual t/a ratio of 1.4 is exceeded. For Type I joints, the strength of joints seems to become asymptotic to the pure infill strength at a slightly higher t/a ratio, especially for large initial normal stresses (i.e. σ_{no} = 1.10 MPa). Furthermore, for Type II joints, the drop in peak shear stress with t/a ratio is much steeper than that of Type I joints. This is naturally because of the considerably higher initial strength of the Type II joints (i.e. at t/a = 0), attributed to the greater frictional resistance associated with the increased asperity angle. At σ_{no} = 1.10 MPa, for Type I joints, the shear strength of the infill joints is significantly less than that of the pure bentonite infill, even at t/a = 1.8. This is because the synthetic soft joints made of gypsum plaster have a smooth finished surface, which at the interface with bentonite can produce an apparent friction angle less than that of pure bentonite. This aspect has also been further discussed elsewhere by Kanji (1974) and de Toledo & de Freitas (1993).

3.3.10 *Drop in peak shear strength*

The normalised strength drop (NSD) has been defined by the reduction in peak shear stress with respect to peak shear stress (τ_p) for clean joints divided by the initial normal stress ($\Delta\tau_p/\sigma_{no}$) (Indraratna et al. 1999). The change in NSD with t/a ratio is plotted in Figure 3.23 for the three levels of σ_{no}. The effect of (a) asperity and infill, and (b) infill only can be observed from these plots, for both Type I and II joints. The rate of normalised strength drop (NSD) is greater at smaller t/a ratios, say, less than 1.4 (i.e. effect of asperity and infill), and it almost becomes asymptotic to the horizontal after the critical t/a ratio is exceeded (i.e. ef-

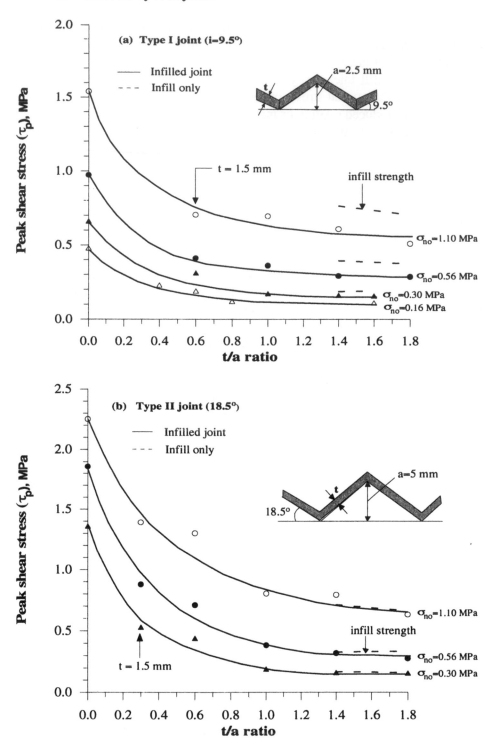

Figure 3.22. Variation of peak shear stress against *t/a* ratio for: a) Type I, and b) Type II infilled joints (Indraratna et al. 1999).

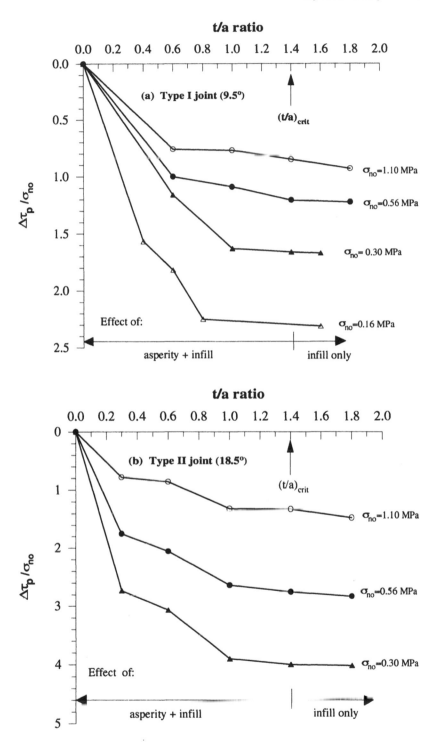

Figure 3.23. Drop in peak shear stress against *t/a* ratio for: a) Type I, and b) Type II infilled joints under various σ_{no} (Indraratna et al. 1999).

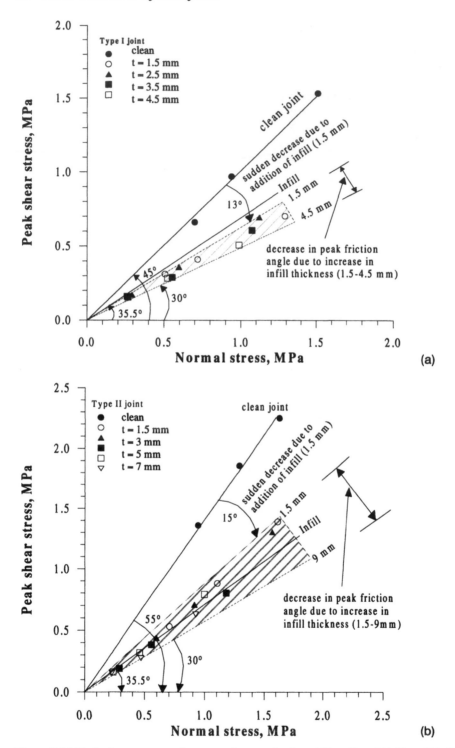

Figure 3.24. Peak shear stress against normal stress plot for: a) Type I, and b) Type II infilled joints (Indraratna et al. 1999).

fect of infill only). The drop in NSD is less marked for higher values of σ_{no}, because of the shearing through asperities, thereby giving a smaller apparent friction angle at elevated normal stresses. As expected, the NSD is greater for the higher asperity angle ($i = 18.5°$) in comparison with Type I joints, at the same initial normal stress level.

3.3.11 *Strength envelope*

The CNS shear strength envelopes for the clean joint, pure infill and for various infill thicknesses are plotted in Figure 3.24 for comparison. From Figure 3.24a, it is clear that the angle of shearing resistance of Type I clean joints sharply decreases after an addition of a thin infill layer of bentonite (1.5 mm). Similar behaviour is observed for the Type II joints as shown in Figure 3.24b. The steep envelopes of the clean joints in Figures 3.24a and 3.24b include the friction angle of planar joints plus the asperity angle (i.e. $\phi_b + i$), as previously described in Patton's model. For the Type I joints where the surface roughness is smaller than Type II joints, the decrease in friction angle with infill thickness is less pronounced. For instance, as thickness (t) is increased from 1.5 mm to 4.5 mm ($t/a =$ 0.6 to 1.8), the reduction in friction angle is not more than 2°. For Type II joints, the reduction in friction angle is more than 10°, as t/a ratio is increased by the same margin.

3.4 SHEAR STRENGTH MODEL FOR INFILLED JOINTS

Lama (1978) established a logarithmic relationship for the prediction of infilled joint shear strength based on laboratory investigations conducted on kaolin filled, rough tension joints of sandstone. The empirical relationship can be represented by the following equation:

$$\tau_p = 7.25 + 0.46\sigma_n - 0.30 \ln (t)\sigma_n^{0.745} \tag{3.1}$$

where τ_p is the shear strength (kN/m^2), σ_n is the normal stress (kN/m^2), and t is the thickness of the infill material (mm). The proposed equation is only applicable for the specific roughness of the joint tested, as the above equation does not contain any terms related to the surface roughness.

Phien-Wej et al. (1990) presented an empirical equation based on laboratory results for the determination of infilled joint strength. They argued that for a low asperity angle, the shear strength envelope would be linear and could become bilinear at higher asperity angles. The joint behaviour was similar to the infill alone when the t/a ratio reached 2. The shear displacement to attain peak strength was greater for higher infill thickness. Based on the above findings, they proposed the following empirical model for the prediction of the infilled joint shear strength:

$$\frac{\tau_p}{\sigma_n} = \frac{\tau_0}{\sigma_n} - \frac{k_1}{\sigma_n}(t/a)\exp[k_2(t/a)] \tag{3.2}$$

where τ_p = shear strength of infilled joint with infill thickness, t; σ_n = normal stress; τ_0 = shear strength of unfilled joint at σ_n; k_1 and k_2 = constants that vary with the surface roughness of joints and applied normal stress.

de Toledo & de Freitas (1993) proposed a general model for the prediction of shear strength of infilled joints for various infill thickness based on the experimental observations as described in Figure 3.25. They adopted the procedure described by Nieto (1974) to describe the infill rock joint shear behaviour as interlocking, interfering and non-interfering. Interlocking occurs when the rock surfaces come in contact; interfering when there is no rock contact, but the strength of the joint is greater than that of the infill alone; non-interfering when the joint behaves as the infill itself. Several researchers have proposed mathematical models for the interfering region (e.g. Ladanyi & Archambault 1977, Papaliangas et al. 1990, Phien-wej et al. 1990). The limit between the interfering and non-interfering regions defines the critical thickness t_{crit} beyond which the joint

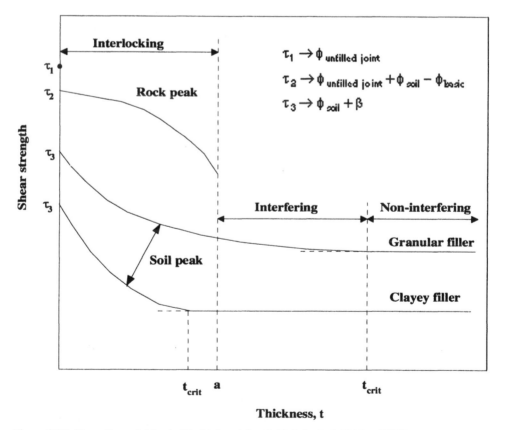

Figure 3.25. Strength model for infilled joints (after de Toledo & de Freitas 1993).

shear behaviour is generally governed by the infill alone. This critical thickness is a function of the grain size of the infill material and the asperity height. Therefore, sands, sandy soils and any material representing granular behaviour tend to have a critical t/a ratio greater than unity. On the other hand, clays present a critical t/a ratio of unity or less.

The joint roughness and size also control the magnitude of the critical t/a ratio. Idealised toothed joints tend to have higher critical thicknesses than tensile fractures; so do small joints as compared with big ones, because the greater the displacement required for rock contact to occur, the easier it is for the infill to achieve its peak strength before rock interference. Experimental evidence shows that a critical t/a ratio of upto 2 is applicable when granular fills are sheared in toothed joints, whereas they may be just above unity when tensile fractures are tested. In the case of clay fills, toothed joints give a critical t/a ratio of unity, which may be as low as 0.60 for tensile fractures. If the t/a ratio is less than unity, rock asperities eventually come in contact with continued shear displacement, generating a second peak shear stress as mentioned by de Toledo & de Freitas (1993). The intercept between the rock peak envelope of an infilled joint for fill thickness tending to zero is lower than the strength of the unfilled joint for a given normal stress. If an unfilled joint is made of weak rock or artificial material, the difference between the rock peak and soil peak may go unnoticed.

Based on the test results obtained for modelled joints, Papaliangas et al. (1993) proposed a model for the prediction of shear strength of infilled joints which incorporates a similar approach to that proposed by Ladanyi & Archambault (1977). The shear strength of a infilled rock joint falls between two limits, τ_{max} the maximum shear strength of the unfilled joints and τ_{min}, the potential minimum shear strength of the system for a critical thickness of infill and which varies with the thickness and type of infill, the roughness of the rock wall and the normal stress. For rough, undulating and strongly steeped joints it is reasonable to assume that τ_{min} equals the shear strength of the infill, but for planar or slightly undulating smooth joints, τ_{min} will be equal to the strength along the interface, which is often lower than the shear strength of the infill. Based on test results, they expressed the peak shear stress as a percentage of stress ratios, as follows:

$$\mu = \mu_{min} + (\mu_{max} - \mu_{min})^n \tag{3.3}$$

where

$$\mu = (\tau/\sigma) \times 100$$

$$\mu_{max} = (\tau_{max}/\sigma) \times 100$$

$$\mu_{min} = (\tau_{min}/\sigma) \times 100$$

$$n = \left[1 - \frac{1}{c}\left(\frac{t}{a}\right)\right]^m$$

for

$$0 \le \frac{t}{a} \le c$$

t = mean thickness of filling material, and a = mean roughness amplitude of the discontinuity.

The constant c is defined as the ratio t/a at which the minimum shear strength is reached, and this depends upon the properties of the filling material, the normal stress and the roughness of the discontinuity surface. The constants c and m are experimentally derived. For the series of tests conducted by Papaliangas et al. (1993), c and m values are considered as 1.5 and 1 for peak, respectively. Similar values were also proposed by Ladanyi & Archambault (1977). For $t/a = 0$, $\mu = \mu_{max}$ which gives the shear strength of the clean joint. For $t/a > c$, μ should be taken equal to μ_{min} which gives the minimum shear strength of the system.

3.5   REMARKS ON INFILLED JOINT BEHAVIOUR

Considering the reported test results, it can be concluded that the shear strength of the joint approaches the strength of the infill at different t/a ratios, depending on the infilling material such as clay or sand. The critical t/a ratio is close to unity for clayey infill (Kanji 1974, Ladanyi & Archambault 1977, Lama 1978, Papaliangas et al. 1990) and is greater than unity for granular infill (Goodman 1970, Papaliangus et al. 1993, Phien-wej et al. 1990). The important factors that have significant influence on the shear behaviour of infilled joints are: joint type, infill type, shear rate, external stiffness, lateral confinement during shearing, and consolidation characteristics. Due to this wide range of controlling factors, it is very important to investigate in detail the shear behaviour of infilled joints under CNS conditions, as almost all the previous tests on infilled joints have been conducted under CNL. The following drawbacks may affect the previous test results when applied to a wide range of practical situations:

– Use of external stiffness as zero, ie. CNL may produce unacceptable results.
– Use of different OCR (over consolidation ratios) for infill during shearing under CNL, i.e. allowing the infill to consolidate at higher normal stress and then applying smaller normal stress during testing. As pointed out by de Toledo & de Freitas (1993), increased OCR makes infill stiffer and thereby providing a greater apparent strength.
– Not all researchers have tested specimens by providing low friction lateral confinement, which is important to represent field situations.

Based on the tests carried out by the authors under CNS condition, the following aspects related to the shear behaviour of infilled joints are highlighted:

– The shear strength of joints can drop considerably due to the addition of a very thin infill layer say, 1.5 mm.

- As the *t/a* ratio approaches 1.4, the change in shear strength becomes insignificant. This *t/a* ratio is defined as the 'critical ratio'.
- Both the asperity angle and infill thickness control the shear behaviour for *t/a* ratio less than unity. However, the infill alone seems to control the shear behaviour, when *t/a* ratio exceeds the critical *t/a* ratio.
- The stress-path plots indicate a reduction in normal stress throughout the test, and an early increase in shear stress followed by a decrement for *t/a* ratio greater than the 'critical' value. However, an increase in normal stress (until the shearing of asperities takes place) is associated with an increase in shear stress over a considerable shear displacement for *t/a* ratios smaller than the 'critical' value.
- The horizontal displacement corresponding to peak shear stress decreases rapidly as infill governs the shear behaviour.
- The peak friction angle for infilled joints is smaller than that of the unfilled joints.

The commonly used models for the prediction of infill joint strength are empirical and are mainly supported by CNL test results (Lama 1978, Phien-Wej et al. 1990, Papaliangas et al. 1993). These models are unable to explain the shearing behaviour under CNS condition, and also they do not address sufficiently the behaviour of infilled joints in comparison with the unfilled joints, in relation to the properties and thickness of infill. Therefore, a new shear strength model that can accommodate both unfilled and infilled joint shear behaviour under the CNS condition is required. The authors have formulated a new approach to predict the shear strength of rock joints, in particular soft rock joints subjected to CNS condition, based on hyperbolic stress-strain relationship. Details of this model are given in Chapter 4 together with the unfilled/clean joints model.

REFERENCES

Barla, G., Forlati, F. & Zaninetti, A. 1985. Shear behaviour of filled discontinuities. *Proc. Int. Symp. on Fundamentals of Rock Joints, Bjorkliden*, pp. 163-172.
Barton, N. & Choubey, V. 1977. The shear strength of rock joints in theory and practice. *Rock Mechanics*, 10: 1-54.
Cheng, F. 1996. A laboratory study of the effects of wall smear and residual drilling fluids on rock socketed pile performance. Ph. D thesis, Monash University Australia.
Cheng, F., Haberfield, C.M. & Seidel, J.P. 1996. Laboratory study of bonding and wall smear in rock socketed piles. *Proc. 7th ANZ Conf. on Geomechanics, Adelaide*, pp. 69-74.
Clark, J.I. & Mayerhof, G.G. 1972. The behaviour of piles driven in clay. An investigation of soil stress and pore pressure as related to soil properties. *Can. Geotech. J.*, 9: 351-373.
de Toledo, P.E.C. & de Freitas, M.H. 1993. Laboratory testing and parameters controlling the shear strength of filled rock joints. *Geotechnique*, 43(1): 1-19.
Ehrle, H. 1990. Model material for shear tests of filled joints. In *Mechanics of Jointed and Faulted Rock*, Rossmanith (ed.), Balkema Publisher, Rotterdam, pp. 371-374.

Eurenius, J. & Fagerstrom, H. 1969. Sampling and testing of soft rock with weak layers. *Geotechnique*, 19(1): 133-139.

Giuseppe, B. 1970. The shear strength of some rocks by laboratory tests. *Proc. 2nd Cong. ISRM, Belgrade*, 2: 3-24.

Goodman, R.E. 1970. The deformability of joints. In determination of the insitu modulus of deformation of rocks. *Special technical publication (ASTM)*, 477: 174-196.

Haque, A. 1999. Shear behaviour of rock joint under constant normal stiffness condition. Ph.D thesis, University of Wollongong, Australia, p. 271.

Indraratna, B., Haque, A. & Aziz, N. 1998. Shear behaviour of idealised infilled joints under constant normal stiffness.*Geotechnique*, 49(3): 331-355.

Kanji, M.A. 1974. Unconventional laboratory tests for the determination of the shear strength of soil-rock contacts. *Proc. 3rd Congr. Int. Soc. Rock Mech., Denver*, 2: 241-247.

Kutter, H.K. & Rautenberg, A. 1979. The residual shear strength of filled joints in rock. *Proc. 4th Int. Congr. Rock Mech., Montreux*, 1: 221-227.

Ladanyi, H.K. & Archambault, G. 1977. Shear strength and deformability of filled indented joints. *Proc. 1st Int. Symp. Geotech. Structural Complex Formations, Capri*, pp. 317-326.

Lama, R.D. 1978. Influence of clay fillings on shear behaviour of joints. *Proc. 3rd Congr. Int. Assoc. Eng. Geol., Madrid*, 2: 27-34.

Papaliangas, T., Hencher, S.R., Lumsden, A.C. & Manolopoulou, S. 1993. The effect of frictional fill thickness on the shear strength of rock discontinuities. *Int. J. Rock Mech. Min. Sci. & Geomech. Abstr.*, 30(2): 81-91.

Pereira, J.P. 1997. Rolling friction and shear behaviour of rock discontinuities filled with sand. *Int. J. Rock Mech. & Min. Sci.*, 34(3-4), paper No. 244.

Pereira, J.P. 1990a. Shear strength of filled discontinuities. In *Rock Joints* Barton & Stephansson (eds), Balkema Publisher, Rotterdam, pp. 283-287.

Pereira, J.P. 1990b. Mechanics of filled discontinuities. In *Mechanics of Jointed and Faulted Rock* Rossmanith (ed.), Balkema Publisher, Rotterdam, pp. 375-380.

Phien-wej, N., Shrestha, U.B. & Rantucci, G. 1990. Effect of infill thickness on shear behaviour of rock joints. In *Rock Joints* Barton & Stephansson (eds), Balkema Publisher, Rotterdam, pp. 289-294.

Sun, W.H., Zheng, T.M. & Li, M.Y. 1981. The mechanical effect of the thickness of the weak intercalary layers. *Proc. Int. Symp. on weak rocks, Tokyo*, 1: 49-54.

Suorineni, F.T. & Tsidzi, K.E.N. 1990. Gemechanical characteristics of a pulverised infilling material of a shear zone. In *Rock Joints* Barton & Stephansson (eds), Balkema Publisher, Rotterdam, pp. 317-321.

Tulinov, R. & Molokov, L. 1971. Role of joint filling material in shear strength of rocks. *ISRM Symp., Nancy*, Paper II-24.

Wanhe, S. Tiemin, Z. & Mingying, L. 1981. The mechanical effect of the thickness of the weak intercalary layers. *Proc. Int. Symp. on Weak Rock, Tokyo*, pp. 49-54.

Xu, S. 1989. The relationship between stress and displacement for rock surfaces in shear. Ph.D thesis, University of London.

CHAPTER 4

Modelling the shear behaviour of rock joints

4.1 INTRODUCTION

A brief overview of the shear strength models for clean/unfilled joints and infilled joints under Constant Normal Stiffness (CNS) have been made in Chapters 2 and 3. The shear strength models based on experimental and analytical methods for determining the shear behaviour of concrete/rock interfaces under CNS have been developed since 1989 (Johnston & Lam 1989, Haberfield & Johnston 1994, Seidel & Haberfield 1995, Johnston et al. 1987). The original objectives of the development of these models were to simulate the side shear behaviour of rock socketed piles. In general, this model deals with the behaviour of two different types of materials having different stiffness values. Thus, the behaviour of rock joint interfaces which have a different behaviour to intact rock or rock mass (Haberfield & Seidel 1998) cannot be predicted accurately using the special concrete/rock interface models.

The use of energy balance theory has been proved to be an effective tool in formulating a shear strength model for rock joints (Ladanyi & Archambault 1970). This method critically divides the total work done involved in shearing a joint into three parts such as:
1. Work done against external force,
2. Work done against dilation of a joint, and
3 Work done against friction.
The shear strength model proposed by Ladanyi & Archambault (1970) is appropriate for predicting the shear behaviour of natural joints which accounts the sliding and shearing mechanisms. However, joints subjected to degradation or plastic deformation tend to slide at a smaller dilation angle, compared to the initial asperity angle (Haberfield & Seidel 1998). In order to explain such joint behaviour, Seidel & Haberfield (1995) pointed out that the work done against external force and internal friction would remain unchanged despite elasticity or plasticity considerations. It is the work done against dilation of a joint which changes with plastic deformation. Thus, for the development of a shear strength model for degradable material (e.g. casting plaster joint), it is essential to incorporate these modifications. Otherwise, the peak shear strength underestimates the actual stresses at failure.

The shear strength models associated with the inclusions of infill are based on empirical relationships (Papaliangas et al. 1993, Phien-Wej et al. 1990, de Toledo & de Freitas 1993). These models neither address the more general non-planar joint behaviour nor account for the normal stiffness of the surrounding rock mass around the joint. Therefore, in this chapter, the development of an analytical model (Indraratna et al. 1999, Haque 1999) for the prediction of both infill and unfilled joint behaviour under CNS condition will be discussed. In view of this, the Fourier transform method as mentioned earlier is used to describe the joint surface before and after testing. Subsequently, the energy balance principle is applied to derive the new strength model under CNS. Using the experimental results and applying the hyperbolic relationships (Duncan & Chang 1970), the drop in strength due to the inclusion of infill in soft joints is quantified under CNS condition.

4.2 EXISTING MODELS BASED ON CNS CONCEPT

Most of the previously developed models are based on the conventional direct shear or CNL condition, and their application to Constant Normal Stiffness (CNS) situations may yield unreliable results. Therefore, attempts made by several researchers to model the shear behaviour of rock joints under CNS are described below.

4.2.1 *Model based on energy balance principles*

Johnston & Lam (1989) developed an analytical method for the shear resistance of concrete/rock interface under CNS condition. Assuming penetration of the micro-asperities of concrete into the rock surface when the contact normal stress exceeds the uniaxial compressive strength, they formulated the following equation for the mobilised cohesion, c_m:

$$C_m = \frac{c_{sl}}{\pi} \cos^{-1}\left(1 - \frac{2\sigma_n}{q_u}\right) \tag{4.1}$$

where c_{sl} = cohesion of rock for asperity sliding, σ_n = actual contact normal stress, and q_u = uniaxial compressive strength.

The equation representing the additional work done in friction due to dilatancy (S_2) as proposed by Ladanyi & Archambault (1970) was modified to incorporate the mobilised cohesive force, and after considering the energy balance principles, the following expression was established to model the average shear stress for sliding:

$$\tau_{sl}^p = (\sigma_{no} + \Delta\sigma_n) \tan\left(i + \phi_{sl}^p\right) +$$

$$+ \frac{\pi c_{sl}}{2\pi \cos^2 i \left(1 - \tan i \tan \phi_{sl}^p\right)} \cos^{-1}\left(1 - \frac{4\tau_{sl}^p \sin i \cos i}{\eta q_u}\right) \qquad (4.2a)$$

where $\Delta\sigma_n = K\Delta y_1$, Δy_1 = dilation caused by shear displacement, K = spring stiffness, i = asperity angle, ϕ_{sl}^p = peak friction angle in sliding, η = interlocking factor, c_{sl} = cohesion of the rock for asperity sliding, and q_u = uniaxial compressive strength.

The average shear stress at shearing to initiate a plane of weakness through asperities was given by:

$$\tau_{sh}^p = (\sigma_{no} + \Delta\sigma_n) \tan\left(\theta_1 + \phi_{sh}^p\right) +$$

$$+ \frac{c_{sh} \tan(i)\eta}{\cos^2 \theta_1 (\tan i + \tan \theta_1)(1 - \tan \theta_1 \tan \phi_{sh}^p)} \qquad (4.2b)$$

where τ_{sh}^p = average shear stress, θ_1 = inclination of shear plane, ϕ_{sh}^p = peak friction angle in shear; σ_{no} = normal stress at initial condition, $\Delta\sigma_n$ = change in normal stress due to dilation, c_{sh} = cohesion for shearing, i = initial asperity angle, and η = interlocking factor.

Once the shear plane is developed and displacement continues along the shear plane, the second term which adds cohesion to the above equation was considered to be zero. Johnston & Lam (1989) also extended the average shear strength expression for subsequent development of shear plane at different inclinations.

The above analytical equations were solved numerically for a given value of joint geometry parameters (i, B, L), where i = initial asperity angle, B = width of specimen, L = length of specimen, and boundary conditions (σ_{no}, K), where σ_{no} = initial normal stress, K = normal stiffness. The method also needs the values of joint strength (q_u), shearing and sliding peak and residual friction angles (ϕ_{sh}^p, ϕ_{sl}^p, ϕ_{sh}^r, ϕ_{sl}^r) and cohesion values (c_{sh}, c_{sl}).

Seidel & Haberfield (1995) extended the energy balance theory proposed by Ladanyi & Archambault (1970) to explain the shear behaviour of more complex joints such as:
– Joints having varying asperity angles, and
– Joints which degrade during shearing.
It was verified that for simple triangular asperities which deform elastically, the Ladanyi & Archambault (1970) formulation based on joint dilation rate was incorrect for rock joints in which high asperity contact stresses may result in significant local elastic deformations. However, joints experiencing plastic deformation cannot be modelled by elastic theory, hence, Ladanyi & Archambault's (1970) approach needed modification. It is proposed by Seidel & Haberfield

(1995) that if a joint dilates at an angle (i_1) less than the initial asperity angle (i_0) i.e. $i_1 < i_0$, then the following equations should be considered based on the energy approach (Fig. 4.1):

S_1 = Component of external work done in dilating a joint against the normal force N, where:

$$S_1 = \frac{N(dy - dp)}{dx} + \frac{Ndp}{dx} = N \tan i \qquad (4.3)$$

where dy = increments of dilation, dp = plastic deformation, N = applied normal force, Ndp = additional work required to increase the internal strain energy of the asperities, dx = shear displacement, and i = initial asperity angle.

S_2 = Component due to additional internal work in friction due to dilatancy = $S\, i_1 \tan \phi_u$, where S = total shear force, i_1 = dilation rate, and ϕ_u = basic friction angle.

S_3 = Component due to internal friction if sample did not change in volume in shear = $N \tan \phi_u$ where ϕ_u is the basic friction angle.

Combining all these three components of 'work done', the following expression was derived to relate shear stress (τ) to normal stress (σ_n):

$$\tau = \frac{\sigma_n (\tan i + \tan \phi_u)}{1 - \tan i_1 \tan \phi_u} \qquad (4.4)$$

where ϕ_u = basic friction angle, i = initial asperity angle, and i_1 = dilation angle.

The validity of the above shear strength equation was verified (Seidel & Haberfield 1995) by conducting CNS shear tests on calcarenite/concrete interfaces containing triangular asperities of angles varying from 5° to 27.5°, under an initial normal stress of 300 kPa. Test results revealed that the proposed model can predict the experimental shear strength results of very closely. In contrast, Patton (1966) model overestimated, and Ladanyi & Archambault (1970) model underestimated the shear strength of joints as shown in Table 2.2.

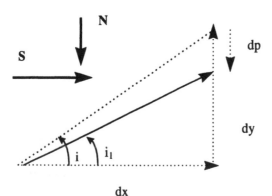

Figure 4.1. Deformation due to inelasticity (after Seidel & Haberfield 1995).

4.2.2 *Mechanistically based model*

A number of rock joint modelling studies have been conducted using triangular asperities for simplicity (e.g. Kodikera & Johnston 1994, Johnston & Lam 1989). Haberfield & Johnston (1994) developed a mechanistically-based model for predicting rough rock joint behaviour under CNS condition. Firstly, they statistically obtained the following roughness parameters to define the interface (Fig. 4.2):

i_m = Mean chord inclination from the horizontal,

i_{sd} = Standard deviation of chord inclination,

h_m = Mean chord height above a horizontal datum, and

h_{sd} = Standard deviation of chord heights.

The shear strength model described by Johnston & Lam (1989) was adopted by Haberfield & Johnston (1994) as a basis for modelling irregular profiles. The distribution of normal force on the individual asperities was determined by the following equation:

$$N_j = n_j^r \cos i_j - s_j \sin i_j \qquad (4.5)$$

where N_j = estimated normal force on asperity j, n_j^r = rebound normal force for asperity j, s_j = shear resistance on asperity j, and i_j = asperity angle of asperity j.

The value of s_j was determined from the following equation:

$$s_j = \frac{c_j L_j}{\cos i_j} + n_j^r \tan \phi_j \qquad (4.6)$$

where c_j and ϕ_j are the cohesion and friction angle for sliding on asperity j.

Considering the relative magnitude of deformations from one asperity to another, the normal force, $\overline{N}_j$, carried by the jth asperity is calculated from the following averaging process:

$$\overline{N}_j = \frac{N_j}{\Sigma N_j} N \qquad (4.7)$$

where N = actual total applied normal force on the joint, and ΣN_j = sum of normal forces on all asperities.

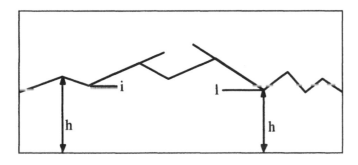

Figure 4.2. Idealized rock joint showing definition of roughness parameters (after Haberfield & Johnston 1994).

If any asperity undergoes shearing, then the normal force carried by the asperity will be different from the above and was calculated by:

$$\overline{N}_j^s = \frac{L_j}{\Sigma L_j} N \tag{4.8}$$

where $\overline{N}_j^s$ is the normal force carried by the jth sheared asperity.

For the intact asperities, the normal force distribution was considered as follows:

$$\overline{N}_j = \frac{N_j}{\Sigma N_j} \left(N - \Sigma \overline{N}_j^s \right) \tag{4.9}$$

where $\Sigma \overline{N}_j^s$ = total normal force carried by the sheared asperities.

The method suggested by Milovic et al. (1970) for the determination of vertical and horizontal displacements of a rigid infinite strip on a finite layer had been employed to predict the displacements for irregular rock interfaces.

4.2.3 *Graphical model*

Saeb & Amadei (1990, 1992) emphasised that constant or variable stiffness boundary conditions are more likely to exist across joint surfaces in-situ rather than constant normal load (CNL) condition. They presented a simple and complete graphical method to predict the shear response of a dilatant rock joint under constant or variable normal stiffness, knowing the behaviour under CNL condition. The method consisted of a series of tests as described below:
- Behaviour of a joint under increasing normal stress with zero shear stress (i.e. normal stress versus joint closure relationship),
- Behaviour of joint under increasing shear stress (i.e. shear stress versus shear displacement relationship and normal displacement versus shear displacement relationship).

A brief summary of the procedure adopted by Saeb & Amadei (1990) is illustrated in Figure 4.3. The method consisted of using curves in Figures 4.3a-c to plot the variation of the joint normal stress (σ_n) versus the joint normal displacement (v). Each curve $u = u_i$ ($i = 0$ to 3) in Figure 4.3d is constructed using the values of σ_n and v at the points of intersection between each line $u = u_i$ and the normal displacement curves in Figure 4.3c. Figures 4.3b and 4.3d can then be used to predict the shear strength of the joint for any load path.

4.2.4 *Analytical model*

Heuze & Barbour (1982) introduced a new model to predict the effect of joint dilation on the behaviour of rock joints. A three-parameter model was introduced to describe the strength envelope below the critical point (σ_c = uniaxial compressive

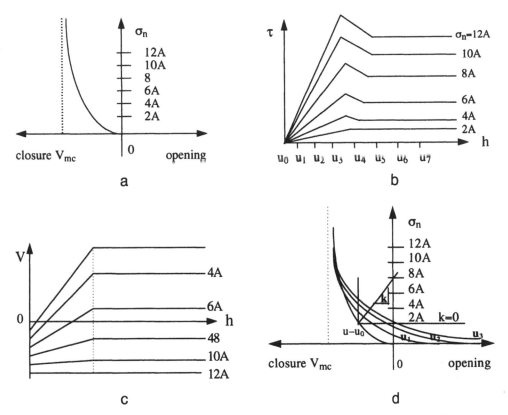

Figure 4.3. Joint response curves. a) Joint closure under increasing normal stress, b) Shear stress versus shear displacement curves for different normal stress, c) Normal displacement versus shear displacement curves at different normal stress, and d) Normal stress versus normal displacement curves for constant stiffness conditions (after Saeb & Amadei 1990). u_0 = initial shear displacement, $u_1, u_2, .. u_7$ = shear displacements, A = arbitrary value of normal stress.

strength) beyond which no dilation was observed. In this model, the peak shear stress (τ_p) is determined by:

$$\tau_p = A\sigma + B\sigma^2 + C\sigma^3 \tag{4.10}$$

where

$$A = \tan \phi_p$$

$$B = \frac{3C_p}{\sigma_c^2} \frac{2(\tan \phi_p - \tan \phi_r)}{\sigma_c}$$

$$C = \frac{-2C_p}{\sigma_c^3} \frac{\tan \phi_p - \tan \phi_r}{\sigma_c^2}$$

The instantaneous dilation angle was found from:

$$\frac{d\tau}{d\sigma} = \tan(\phi_r + \delta) \tag{4.11}$$

where $\delta = \tan^{-1}(A + 2B\sigma + 3C\sigma^2) - \phi_r$.

When $\sigma > \sigma_c$ the peak strength was simply given by $\tau_p = C_p + \sigma \tan\phi_r$, and in the residual range by $\tau_r = \sigma \tan\phi_r$.

However, the normal stress used in the above equations needs to be estimated for dilatant joints before predicting the shear strength of the joint. Based on the conceptual model described in Figure 4.4, Heuze & Barbour (1982) presented the following incremental normal stress equation:

$$\Delta\sigma = \tan\delta \frac{KN \cdot KNEFF}{KN + KNEFF} \Delta u \tag{4.12}$$

where KNEFF = stiffness of the adjacent structure, KN = normal stiffness of the joint itself, Δu = shear displacement along joint, and Δv = normal joint displacement.

Skinas et al. (1990) described a joint model based on CNS condition adopting the mobilised dilation concept of the JRC-JCS model of Barton et al. (1985). The change in dilation with the change in shear displacement (Fig. 4.5) was expressed as:

$$\Delta v = \Delta u \cdot \tan d_n \text{ (mob)} \tag{4.13}$$

where

$$d_n \text{ (mob)} = \frac{1}{M} \text{JRC (mob)} \log\left(\frac{JCS}{\sigma_n}\right)$$

in which M = damage coefficient, JRC (mob) = mobilised joint roughness coefficient, and JCS = joint wall compressive strength.

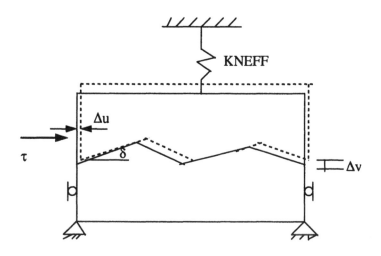

Figure 4.4. Conceptual model of a dilatant joint undergoing shear (after Heuze & Barbour 1982).

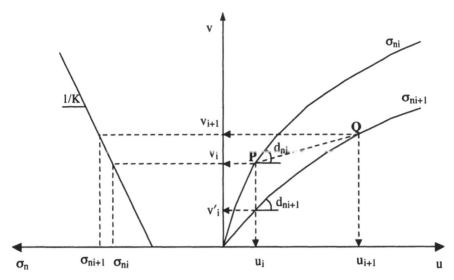

Figure 4.5. Calculation procedure for modelling dilation behaviour under CNS (after Skinas et al. 1990).

The dilation of any point, Q on the dilation-shear displacement curve (Fig. 4.5) and corresponding normal stress can be computed by:

$$v_{i+1} = v'_i + (u_{i+1} - u_i) \tan d_{ni+1} \tag{4.14}$$

and

$$\sigma_{ni+1} = \sigma_{ni} + K (v_{i+1} - v_i) \tag{4.15}$$

Taking into account of the mobilised dilation, the above dilation equation could be rearranged in the following form:

$$v_{i+1} = v'_i + (u_{i+1} - u_i) \tan \left[\frac{1}{M} JRC_m^{ui} \cdot \log \left(\frac{JCS}{\sigma_{ni+1}} \right) \right] \tag{4.16}$$

where subscript 'm' stands for mobilisation, and u_i = any shear displacement.

Accordingly, the normal stress increment $\Delta\sigma$ was calculated by knowing the dilation, v_{i+1} as:

$$\Delta\sigma = K . v_{i+1} / A$$

where A is the joint total area.

Finally, the mobilised shear stress for any stage of shearing was obtained from the following equation:

$$\tau_{(mob)} = \sigma_{ni+1} \tan \left[JRC_m^{u_{i+1}} \log \left(\frac{JCS}{\sigma_{ni+1}} \right) + \phi_r \right] \tag{4.17}$$

where ϕ_r = residual angle of friction.

4.3 REQUIREMENT OF A NEW MODEL

The analytical models developed by Johnston & Lam (1989) and Haberfield & Johnston (1994) are applicable to concrete/rock interface problems in practice. The application of these models to more general rock/rock interface problems (e.g. reinforced slope stability analysis, joints on the roof of an underground excavation) may lead to unrealistic results. Therefore, the effect of a wearing mechanism in the development of a shear strength model needs to be incorporated as pointed out by Seidel & Haberfield (1995).

Saeb & Amadei's (1990, 1992) graphical method for the determination of unfilled joint shear strength under CNS from the data obtained under CNL seems simple to apply, but it is a complicated process due to the wide range of CNL test data. It has been already demonstrated that only a few CNS tests are sufficient to explain the joint behaviour controlled by the stiffness of the surrounding rock mass. The application of energy balance principles in describing the shear behaviour of rock joints can produce acceptable results (Ladanyi & Archambault 1970, Seidel & Haberfield 1995). Skinas et al. (1990) formulation is difficult to use as it is based on the mobilised JRC which again is difficult to obtain with the progress of shearing.

Therefore, in this chapter, the authors attempt to improve the understanding of shear behaviour of rock/rock interfaces by introducing a simple but accurate method of predicting joint dilation under CNS using Fourier Transform method. Thereafter, using energy balance principles, a more reliable form of shear strength equation is developed. This model will be extended further to incorporate the effect of infill material on joint shear behaviour. Detail formulations of the authors' model are discussed in the subsequent sections.

Laboratory CNS and CNL direct shear behaviour are also simulated numerically using Universal Distinct Element Code (UDEC). Finally, the numerical predictions are compared with the analytical and laboratory results to some extent.

4.4 NEW SHEAR STRENGTH MODEL FOR SOFT ROCK JOINTS

4.4.1 *Application of Fourier transform method for predicting unfilled joint dilation*

The Fourier transform method (Spiegel 1974) has been adopted to characterise the joint surfaces before and after shearing of specimens. It has been used as an effective tool for characterisation of metal surfaces in mechanical engineering applications. However, the successful application of the Fourier series in the field of rock mechanics is not yet established (Indraratna et al. 1995). Fourier series can be used precisely to define any continuous function $f(x)$ which is integrable along the period 2π, and has an integrable derivative at some interval (a, b). The fol-

lowing form of Fourier series is used in this study to characterise the joint profile
for a prescribed period, $T = b - a$:

$$\delta_v(h) = \frac{a_0}{2} + \sum_{n=1}^{\infty} \left[a_n \cos\left(2\pi n h / T\right) + b_n \sin\left(2\pi n h / T\right) \right] \tag{4.18}$$

where

$$a_n = \frac{2}{T} \int_a^b f(x) \cos\frac{2\pi n x}{T} \, dx \tag{4.18a}$$

and

$$b_n = \frac{2}{T} \int_a^b f(x) \sin\frac{2\pi n x}{T} \, dx \tag{4.18b}$$

Fourier series can be used to match the exact joint dilation with horizontal displacement (Fig. 4.6), where the Fourier coefficients a_n and b_n can be determined
based on the experimental data. Given a large array of data points, the integral
Expressions 4.18a and 4.18b can be simplified to the following summation series,
by subdividing the dilation versus horizontal displacement plot into m equal parts
(Fig. 4.6), and then using the rectangular rule for simplicity:

$$a_n \approx \frac{2}{m} \sum_{k=0}^{m-1} y_k \cos\frac{2k\pi}{m} n \tag{4.19a}$$

$$b_n \approx \frac{2}{m} \sum_{k=0}^{m-1} y_k \sin\frac{2k\pi}{m} n \tag{4.19b}$$

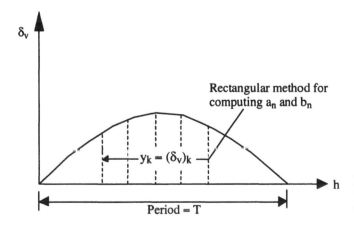

Rectangular method for
computing a_n and b_n

$y_k = (\delta_v)_k$

Period = T

Figure 4.6. Typical dilation
behaviour of saw-tooth joint
under CNS.

For the current (two dimensional roughness) test data, the Fourier coefficients are calculated for Types I, II, III and natural (field) joints before and after shearing under different initial normal stresses (σ_{no}), as given in Tables 4.1 and 4.2 for $T = 30$ mm.

4.4.2 Prediction of normal stress with horizontal displacement

Once the joint dilation, $\delta_v(h)$ with horizontal displacement (h) under a given initial normal stress (σ_{no}) is fitted to a Fourier series (Eq. 4.18), the variation of

Table 4.1. Fourier coefficients (4 harmonics) for Types I, II and III joints for various initial normal stresses.

Joint type	Initial normal stress (σ_{no}), MPa	a_0	a_1	a_2	a_3	b_n
Type I	Original profile	2.5	−1.036	0	−0.137	0.00
	0.16	1.655	−0.657	−0.053	−0.035	0.00
	0.30	1.727	−0.641	−0.115	−0.047	0.00
	0.56	1.437	−0.502	−0.083	−0.053	0.00
	1.10	1.38	−0.414	−0.115	−0.057	0.00
	1.63	0.968	−0.347	−0.059	−0.027	0.00
Type II	Original profile	5	−2.07	0	−0.28	0.00
	0.16	3.76	−1.37	−0.22	−0.11	0.00
	0.30	3.38	−1.20	−0.26	−0.10	0.00
	0.56	2.94	−1.03	−0.21	−0.11	0.00
	1.10	2.07	−0.78	−0.14	−0.055	0.00
	1.63	1.50	−0.56	−0.123	−0.032	0.00
	2.43	0.55	−0.095	−0.115	−0.013	0.00
Type III	Original profile	7.50	−3.10	0.00	−0.40	0.00
	0.16	3.51	−0.69	−0.54	−0.19	0.00
	0.30	3.64	−1.06	−0.46	−0.10	0.00
	0.56	2.64	−0.59	−0.47	−0.14	0.00
	1.10	1.43	−0.28	−0.29	−0.03	0.00
	1.63	1.4	−0.35	−0.22	−0.07	0.00
	2.43	0.97	−0.19	−0.21	−0.05	0.00

Table 4.2. Fourier coefficients for field joints for various σ_{no} values under CNS.

σ_{no}, MPa	a_0	a_1	a_2	a_3
0.56	0.26	−0.18	−0.05	−0.02
1.10	0.14	−0.07	−0.04	−0.005
1.63	−0.15	−0.04	0.02	0.016
2.16	−0.32	0.04	0.01	0.025
2.69	−0.51	0.06	0.05	0.035

normal stress under constant normal stiffness (k_n) can be determined by Equation 4.20:

$$\sigma_n(h) = \sigma_{no} + \Delta\sigma_{n.h} = \sigma_{no} + \frac{k_n.\delta_v(h)}{A} \qquad (4.20)$$

where $\sigma_n(h)$ = normal stress at any horizontal displacement, h; σ_{no} = initial normal stress; $\Delta\sigma_{n.h}$ = incremental normal stress at any shear displacement, h = $[k_n.\delta_v(h)/A]$; k_n = normal stiffness; $\delta_v(h)$ = dilation corresponding to horizontal displacement given by Equation 4.18; A = joint surface area.

4.4.3 *Prediction of shear stress with horizontal displacement*

Newland & Alley (1957) proposed that the shear resistance of rock joints can be explained using Equation 4.21 based on tests conducted on granular materials such as sand.

$$\tau = \sigma_n \tan(\phi_b + i) \qquad (4.21)$$

where σ_n = applied normal stress, and i = inclination of the saw-tooth asperity.

Patton (1966) used the above equation for explaining the shear resistance of regular saw-tooth joints produced from casting plaster. He then mentioned that the shear strength of the saw-tooth shaped joints can be explained well by Equation 4.21 for a low range of applied normal stress where the asperity degradation during shearing is negligible. However, when the joints are sheared under a higher normal stress range, the chance of shearing the asperities is greater (Fig. 4.7) and the shear behaviour can no longer be explained by Equation 4.21. Patton

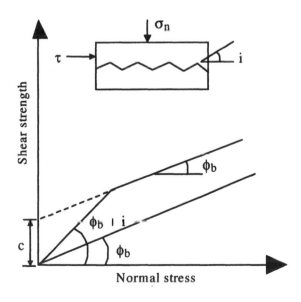

Figure 4.7. Strength envelope for saw-tooth asperity of inclination, *i* (after Patton 1966).

(1966) observed that the shear resistance under high normal stress levels can be explained by Equation 4.22 as follows:

$$\tau = c + \sigma_n \tan(\phi_b) \tag{4.22}$$

where ϕ_b = basic friction angle which can be determined by testing flat joints, and c = cohesion of joint.

Ladanyi & Archambault (1970) explained Patton's (1966) formulation by considering the energy principles described by Rowe et al. (1964) on regular triangular asperities as:

$$S = S_1 + S_2 + S_3 \tag{4.23}$$

where S = total shear resistance, $S_1 = \sigma_n \tan(i)$ = component of external work done in dilating against external stress σ_n, $S_2 = S \tan(i) \tan(\phi_b)$ = component of additional work done against internal friction due to dilatancy, and $S_3 = \sigma_n \tan(\phi_b)$ = component of work done in friction if the sample did not change volume during shearing.

Combining the force components of the work done, Ladanyi & Archambault (1970) derived a shear strength expression which is identical to that obtained by Patton (Eq. 4.22) for the peak sliding resistance of sawtooth profiles with inclination, i, at normal stress levels less than the transition stress. Ladanyi & Archambault (1970) analysis is based on the assumption of rigid asperities, where the dilation rate is equal to the tangent of the asperity angle, i. However, for the asperities which undergo degradation, the rate of dilation changes with shear displacement, and the above formulation no longer remains valid.

Once the dilation and normal stress responses are predicted using Equations 4.18 and 4.20 for an initial asperity angle (i) and joint dilation rate (i_h), the shear stress response with the horizontal displacement can be calculated by a modified form of Patton equation (Eq. 4.24) as given below:

$$\tau_{(h)} = (\sigma_{no} + \Delta\sigma_{n.h}) \tan(\phi_b + i_h) = (\sigma_{no} + \Delta\sigma_{n.h}) \left(\frac{\tan(\phi_b) + \tan(i_h)}{1 - \tan(\phi_b)\tan(i_h)} \right)$$

$$\tag{4.24}$$

where σ_{no} = initial normal stress; $\Delta\sigma_{n.h}$ = incremental normal stress at any shear dispalcement, h is given in Equation 4.20; ϕ_b = basic friction angle; i_h = inclination of the tangent to the dilatancy curve at any horizontal displacement, h.

Seidel & Haberfield (1995) explained the energy principle involved during the shearing of a joint with plastic deformation. It is evident that the work done involved in dilating a joint against a given normal stress is S_1 and the work done in friction is S_3, if the volume change is unaffected by the elasto-plastic deformations. The work done in friction due to dilatancy, S_2 will only change as the relative dilatancy is reduced due to the degradation of asperities. Therefore, the term,

S_2 is reduced if the current dilation angle $i_h < i$. On this basis, the expression for the component of additional work done in friction due to dilatancy (S_2) can be represented by:

$$S_2 = S \tan (i_h) \tan (\phi_b) \tag{4.25}$$

In order to satisfy the energy balance principle, the component of the work done term $S_1 = \sigma_{n.h} \tan (i_h)$ needs to be replaced by $S_1 = \sigma_{n.h} \tan (i)$ in Equation 4.24. Consequently, the shear behaviour depicted by Equation 4.24 will result in an un deroatimation of the shear stress of the joint, unless it is modified and rewritten in the following form:

$$\tau_h = \sigma_{n.h} \left(\frac{\tan (\phi_b) + \tan (i)}{1 - \tan (\phi_b) \tan (i_h)} \right) \tag{4.26a}$$

Replacing $\sigma_{n.h}$ in Equation 4.24 and expressing in terms of Fourier equations, Equation 4.26a can be written as:

$$\tau_h = \left[\sigma_{no} + \frac{k_n}{A} \left(\frac{a_0}{2} + \sum_1^n \left(a_n \cos \frac{2\pi nh}{T} + b_n \sin \frac{2\pi nh}{T} \right) \right) \right]$$

$$\left(\frac{\tan (\phi_b) + \tan (i)}{1 - \tan (\phi_b) \tan (i_h)} \right) \tag{4.26b}$$

where i = initial asperity angle; i_h = inclination of the tangent to the dilatancy curve at any horizontal displacement, h; ϕ_b = basic friction angle; k_n = external normal stiffness; A = joint surface area; σ_{no} = initial normal stress; T = period; h = shear displacement, a_n; b_n = Fourier coefficients.

To obtain the maximum shear stress (τ_p), Equation 4.26b needs to be differentiated with respect to horizontal displacement, h such that $d\tau(h)/dh = 0$. In order to achieve this, Equation 4.18 is differentiated first with respect to h and the following equation is obtained:

$$\tan (i_h) = \frac{d(\delta_v)_h}{dh} = -\frac{2\pi n}{T} \left[\sum_1^n a_n \sin \left(\frac{2\pi nh}{T} \right) - \sum_1^n b_n \cos \left(\frac{2\pi nh}{T} \right) \right] \tag{4.27}$$

At $h = 0$, the slope of the tangent to the dilatancy curve becomes $\tan (i_h) = \tan (i)$, where i = initial asperity angle of the sawtooth asperity. Substituting the value of $\tan (i_h)$ in Equation 4.26b and differentiating with respect to h, the following expression is obtained for regular triangular sawtooth asperities for $b_n = 0$ and $d\tau_h/dh = 0$ conditions:

$$\frac{d\tau_h}{dh} = \frac{k_n}{A}\sum_1^n na_n \sin\left(\frac{2\pi nh}{T}\right)\left[1 + \frac{2\pi}{T}\tan(\phi_b)\sum_1^n na_n \sin\left(\frac{2\pi nh}{T}\right)\right] +$$

$$\left(\frac{2\pi}{T}\tan\phi_b \sum n^2 a_n \cos\frac{2\pi nh}{T}\right)\left[\sigma_{no} + \frac{k_n}{A}\left(\frac{a_0}{2} + \sum_1^n a_n \cos\frac{2\pi nh}{T}\right)\right] = 0$$

(4.28)

As both $\sigma_n(h)$ and $i(h)$ are continuous functions, the solution for 'peak' shear stress (τ_p) always exists, and it can be obtained either graphically (Fig. 4.8) or numerically using a computer subroutine. The horizontal displacement corresponding to peak shear stress ($h_{\tau p}$) can be obtained by solving Equation 4.28 numerically. This value is then used to calculate the dilation at peak shear stress using Equation 4.18. Subsequently, Equations 4.20, 4.26b and 4.27 are employed to obtain shear stress of unfilled joints based on the energy balance principles. Once the peak shear strength of the unfilled joint is determined, the shear strength of the infilled joint can be determined using the methodology described in the subsequent sections.

The Fourier coefficients for the saw-tooth and natural joint surfaces subjected to shearing under various initial normal stresses (σ_{no}) are plotted in Figures 4.9 and 4.10, respectively. The plots reveal that the Fourier coefficients having subscripts 2, 3, 4, ...n have a very small effect on the dilation behaviour of regular triangular joints, hence, on the normal stress and shear stress responses. Therefore, a more simplified solution of Equation 4.28 can be achieved by considering only one harmonic, i.e. $n = 1$. Considering the appropriate Fourier coefficients in Equation 4.28 and equating the $d\tau/dh = 0$, the solution for $h_{\tau peak}$, horizontal displacement corresponding to peak shear stress (τ_{peak}) is given by Equation 4.29 after simplification.

$$\tan\left(\frac{\pi h_{\tau_p}}{T}\right) = \frac{-1 \pm \sqrt{1 - \left(\frac{\pi \tan\phi_b}{T}\right)^2 \left[4a_1^2 - \left(\frac{2\sigma_{no}}{k_n/A} + a_0\right)^2\right]}}{\frac{\pi \tan\phi_b}{T}\left[2a_1 - \left(\frac{2\sigma_{no}}{k_n/A} + a_0\right)\right]}$$

(4.29)

or

$$h_{\tau peak} = \left(\frac{T}{\pi}\right)\tan^{-1}\left[\frac{-1 \pm \sqrt{1 - \left(\frac{\pi \tan\phi_b}{T}\right)^2 \left[4a_1^2 - \left(\frac{2\sigma_{no}}{k_n/A} + a_0\right)^2\right]}}{\frac{\pi \tan\phi_b}{T}\left[2a_1 - \left(\frac{2\sigma_{no}}{k_n/A} + a_0\right)\right]}\right]$$

(4.30)

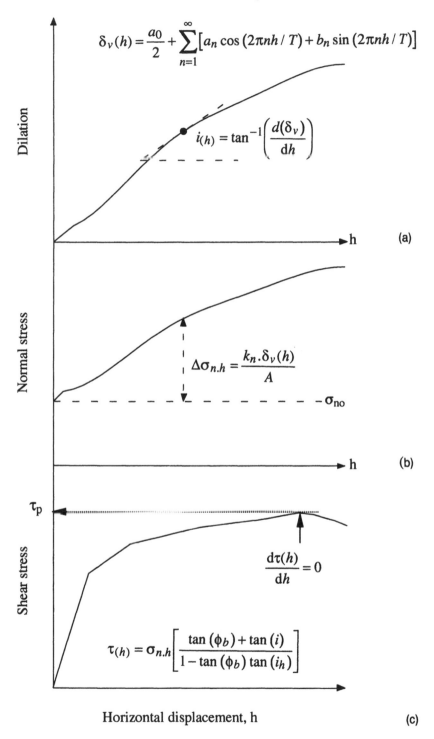

Figure 4.8. Graphical representation of prediction of unfilled joint shear strength (Indraratna et al. 1999).

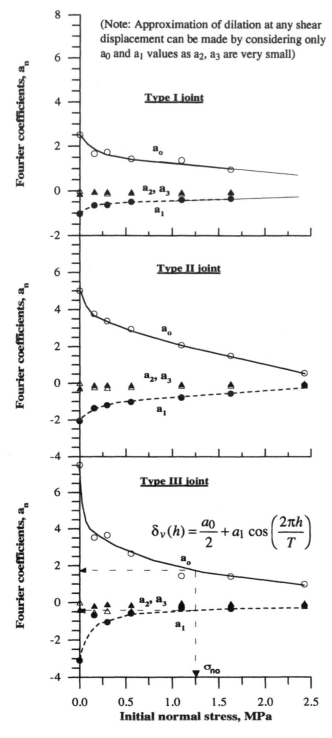

Figure 4.9. Method of approximation of joint dilation at any shear displacement.

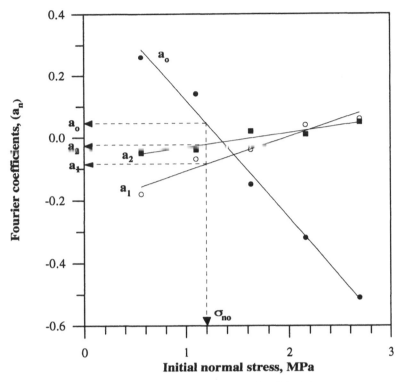

Figure 4.10. Approximation of Fourier coefficients for field joints at any σ_{no}.

By substituting the values of Fourier coefficients (a_0, a_1), normal stiffness (k_n) in N/mm, joint surface area (A) in mm^2, basic friction angle (ϕ_b) and period $T = 30$ mm for regular asperities in Equation 4.30, the value of the horizontal displacement at peak shear stress for a given initial normal stress (σ_{no}) can be calculated. The dilation at $h_{\tau peak}$ can be obtained from Equation 4.18, and the normal stress using Equation 4.20. The dilation angle at peak shear stress ($i_{h_{\tau p}}$) can be obtained from Equation 4.27. Finally, expressing all variables in terms of $h_{\tau peak}$, the following simplified form of the shear strength equation is deduced:

$$\tau_p = \left[\sigma_{no} + \frac{k_n}{A} \left(\frac{a_0}{2} + a_1 \cos \frac{2\pi h_{\tau_p}}{T} \right) \right] \left(\frac{\tan \phi_b + \tan i}{1 - \tan \phi_b \tan i_{h_{\tau p}}} \right) \tag{4.31}$$

where $h_{\tau p}$ and $i_{\tau p}$ are horizontal displacement and dilation angle corresponding to peak shear stress, k_n = normal stiffness, i = initial asperity angle, σ_{no} = initial normal stress, ϕ_b = basic friction angle, A = joint surface area, a_0, a_1 = Fourier coefficients, and T = period.

The accurate determination of shear behaviour of unfilled joints under CNS is accomplished using a computer code (see Appendix), which incorporates the Fourier coefficients for a wide range of harmonics. Subsequently, the peak shear

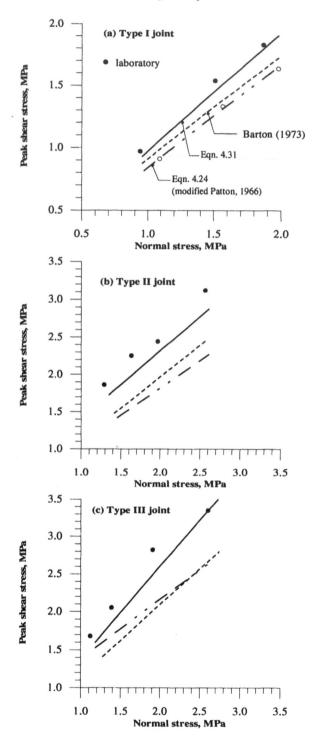

Figure 4.11. Comparison of model predictions with laboratory results for Type I, II and III joints.

Table 4.3. Peak shear stress for Type I, II and III joints predicted by different methods.

Joint type	Initial normal stress σ_{no}, MPa	Peak shear stress, Mpa				Normal stress corresponding to peak shear stress, Mpa	
		Numerical solution (Eq. 4.26b)	Simplified solution (Eq. 4.31)	Barton (1973)	Export value	Numerical solution	Simplified solution
Type I	0.56	1.04	1.05	0.98	0.97	1.07	1.09
	1.10	1.48	1.52	1.38	1.54	1.55	1.57
	1.63	1.89	1.91	1.73	1.83	1.97	1.98
Type II	0.56	1.90	1.96	1.63	1.86	1.60	1.62
	1.10	2.15	2.20	1.86	2.25	1.86	1.85
	1.63	2.45	2.53	2.11	2.44	2.16	2.17
	2.43	2.98	2.87	2.45	3.12	2.56	2.58
Type III	0.56	1.97	1.84	1.52	1.68	1.25	1.39
	1.10	2.06	1.98	1.75	2.05	1.50	1.53
	1.63	2.80	2.71	2.16	2.82	1.95	2.07
	2.43	3.72	3.50	2.76	3.35	2.65	2.71

stress calculated through this process is utilised to determine the strength of in-filled joints as discussed further in Section 4.5. The performance of the simplified shear strength model (Eq. 4.31) over the numerical predictions using Equation 4.26 and well known Barton (1973) model are summarised in Table 4.3. Joint Roughness Coefficients (JRC) for saw-tooth joints are considered as half of the initial asperity angle (i.e. $i_0/2$) on the basis of a simplified method suggested by Maksimovic (1996). It is of interest to note that Equation 4.31 provides a closer prediction of the shear strength in relation to Equation 4.26b (Fig. 4.11). A comparison of the predicted shear strength values by different models (e.g. Eqs 4.24 and 4.31, and Barton 1973) with the laboratory results is presented graphically in Figure 4.11. It is observed that Barton's model underpredicts the shear strength of soft joints under CNS, although it is still adequate for describing the shear behaviour of rock joints under CNL condition (Fig. 4.11).

4.5 EFFECT OF INFILL ON THE SHEAR STRENGTH OF JOINT

4.5.1 *Hyperbolic modelling of strength drop associated with infill thickness*

The effect of infill on the drop of shear strength of rock joints is generally recognised (Ladanyi & Archambault 1977, Barla et al. 1985). A typical strength drop ($\Delta\tau_p$) relative to the peak shear stress (τ_p) for clean joints due to the inclusion of various infill thickness between the interfaces is shown in Figure 4.12a. The drop in strength is normalised by the initial normal stress (σ_{no}) to obtain the normalised strength drop (NSD). The change in NSD of infilled joints can be simulated using a hyperbolic fit. For this purpose, the simplified methodology proposed by Duncan & Chang (1970) has been adopted, as illustrated graphically in Figure 4.12. The relationship between NSD and t/a ratio in Figure 4.12b can be expressed by the following algebraic function:

$$\mathrm{NSD} = \frac{t/a}{\alpha\,(t/a) + \beta}$$

(4.32)

where NSD $= (\Delta\tau/\sigma_{no})$, α and β = constants depending on σ_{no} and surface roughness.

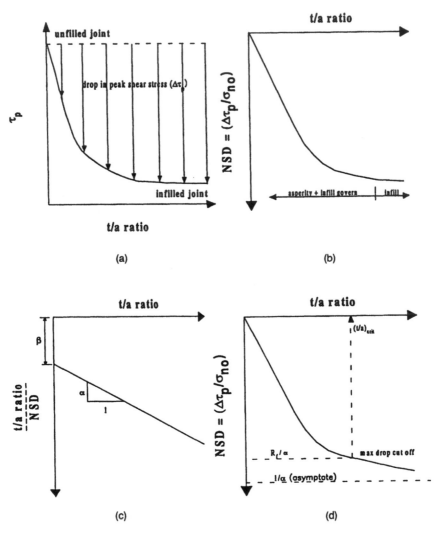

Figure 4.12. Formulation of hyperbolic model for the prediction of drop in peak shear stress due to infill (Indraratna et al. 1999).

The axes of Figure 4.12b can be transformed (Fig. 4.12c) to establish a linear relationship to give the following equation:

$$\frac{t/a}{\text{NSD}} = \beta + \alpha\,(t/a) \tag{4.33}$$

The magnitudes of α and β are obtained by plotting the relationship between t/a and NSD as shown in Figure 4.12c.

Figure 4.12d shows that as the value of t/a ratio reaches a certain limit, further decrease in NSD becomes almost insignificant, or conversely, the infill controls the shear behaviour. The hyperbolic relationship gives $1/\alpha$ as the asymptote, where α is introduced in Equation 4.32. For all the tests, it is noted that the maximum drop in peak shear stress is reached before it becomes asymptotic to $1/\alpha$ (Fig. 4.12d), where reduction factor, R_f varies from 0.80 to 0.90.

4.5.2 *Shear strength relationship between unfilled and infilled joints*

The peak shear strength of an infilled joint under CNS condition can be calculated according to Equation 4.34, once the strength of unfilled joints is known at a given initial normal stress (σ_{no}) for a particular joint profile.

$$\left(\tau_p\right)_{\text{infilled}} = \left(\tau_p\right)_{\text{unfilled}} - \Delta\tau_p \tag{4.34}$$

where $\Delta\tau_p = \sigma_{no} \times \text{NSD}$ as defined by Equation 4.32.

Equation 4.34 can be written in the following simplified form, considering all the parameters influencing the shear behaviour under CNS:

$$\left(\tau_p\right)_{\text{infilled}} = \left[\sigma_{no} + \frac{k_n}{A}\left(\frac{a_0}{2} + a_1\cos\frac{2\pi h_{\tau_p}}{T}\right)\right] \times$$
$$\left(\frac{\tan\phi_b + \tan i}{1 - \tan i\,\tan i_{h_{\tau_p}}}\right) - \sigma_{no}\left(\frac{t/a}{\alpha \times t/a + \beta}\right) \tag{4.35}$$

where h_{τ_p} and i_{τ_p} = horizontal displacement and dilation angle corresponding to peak shear stress, k_n = normal stiffness, i = initial asperity angle, σ_{no} = initial normal stress, ϕ_b = basic friction angle, A = joint surface area, a_0 and a_1 = Fourier coefficients, T = period, t/a = infill thickness to asperity height ratio, α and β = hyperbolic constants as defined in Equation 4.32.

4.5.3 *Determination of hyperbolic constants*

The test results obtained for Type I and II infilled joints are plotted in Figure 4.13 to determine the values of the empirical constants (α and β), and also to compare the accuracy of the hyperbolic fit according to Equation 4.33. It is observed that for both Type I and Type II infilled joints, Equation 4.33 can predict the NSD accurately for the experimental data obtained for the three stress levels of σ_{no}. The values of α and β are given in tabular form on Figure 4.13. It is important to note that these hyperbolic parameters for Type I joints are almost double those of Type II joints, except for the β value at $\sigma_{no} = 1.10$ MPa for Type II joints. Considering a reduction factor (R_f) of 0.85 and the critical t/a ratio of 1.40, the NSD

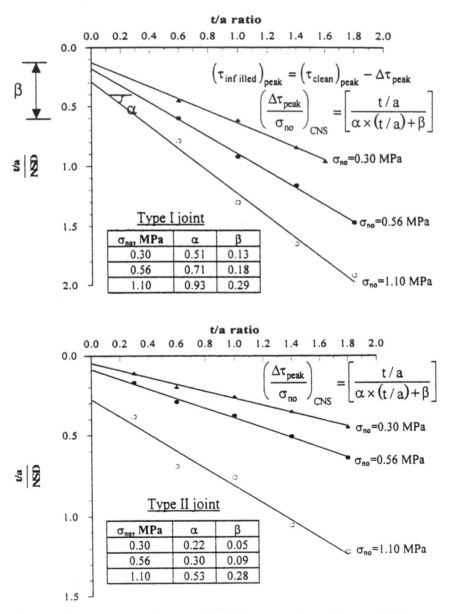

Figure 4.13. Determination of hyperbolic constants α and β for Type I and II infilled joints.

($\Delta\tau_p/\sigma_{no}$) values for a wide range of t/a ratios and σ_{no} values are determined using Equation 4.32. A reduction factor (R_f) is determined from careful laboratory testing, which restricts the 'critical t/a ratio' beyond which the NSD remains almost constant. As indicated in Figure 4.14, the drop in shear strength as a function of the t/a ratio can be predicted well for a given initial normal stress, σ_{no}.

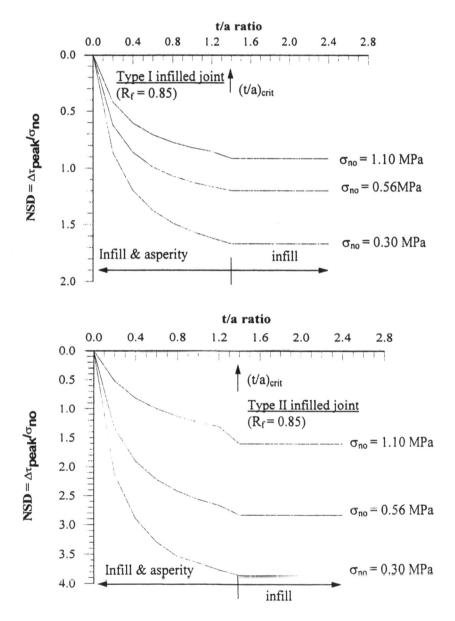

Figure 4.14. Normalised drop in peak shear stress (NSD) for Type I and II infilled joints, based on hyperbolic model predictions.

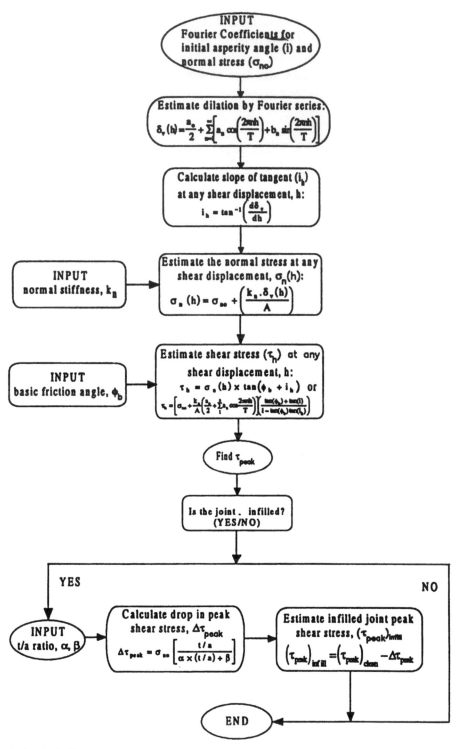

Figure 4.15. Flow chart for the calculation of the peak shear strength of unfilled and filled joint.

4.6 DEVELOPMENT OF A COMPUTER CODE

In order to obtain the Fourier coefficients and an accurate solution of the shear strength equation, a computer code was written according to the flowchart shown in Figure 4.15. Firstly, the program determines the Fourier coefficients from the input data file of dilation behaviour observed in the laboratory. Then it calculates the dilation angle depending upon the horizontal displacement. The correction for normal compliance is made before calculating the normal stress according to the observed value for tests conducted for $\sigma_{no} < 0.56$ MPa. The computer program invokes the required correction data file automatically once the normal stress is less than 0.56 MPa. The normal stress is then calculated using in Equation 4.20. The program executes the stress variations with horizontal displacements using Equations 4.24 (Patton type) and 4.26b (energy balance). Both these equations are employed to compare the predicted results of individual models with laboratory test data. Finally, the shear strength of infilled joints is calculated using Equation 4.34, where the unfilled joint strength is obtained from Equation 4.26b. The infill joint strength in relation to various infill thickness to asperity height ratio (t/a) can be calculated by prescribing the required hyperbolic constants as input parameters.

4.7 COMPARISON BETWEEN PREDICTED AND EXPERIMENTAL RESULTS

The mathematical model is employed to predict the shear behaviour of saw-tooth and natural joints using Equations 4.18 to 4.35 under the CNS condition. A comparative discussion of the predicted and experimental results in relation to dilation, normal stress and shear stress for unfilled saw-tooth and natural (field) joints, as well as for infilled joints is given below.

4.7.1 *Dilation*

The dilation of the joint with shear displacement for various initial normal stresses (σ_{no}) is calculated using the Fourier transform method (Eq. 4.18) for Type I, II and III joints. It is observed that the predicted dilation curves fit the experimental results very closely (Figs 4.16a, 4.17a, and 4.18a). The Fourier method is also extended for natural joints assuming that dilation is continuous for a given period ($T = 10$ to 30 mm), and it was found that the dilation could be predicted to match the experimental results within acceptable limits (Fig. 4.19c).

4.7.2 *Normal stress*

The change in normal stress with horizontal displacement is calculated using

Equation 4.20 for all types of joints. The spring behaviour was corrected for normal compliance for $\sigma_{no} < 0.56$ MPa. Using the predicted dilation, the normal stress response with horizontal displacement is calculated for various values of σ_{no}. The predicted normal stress with horizontal displacement for Type I, II and III and natural joints is shown in Figures 4.16b, 4.17b, 4.18b and 4.19b. It is verified that Equation 4.20 can predict the normal stress to be in close agreement with experimental results.

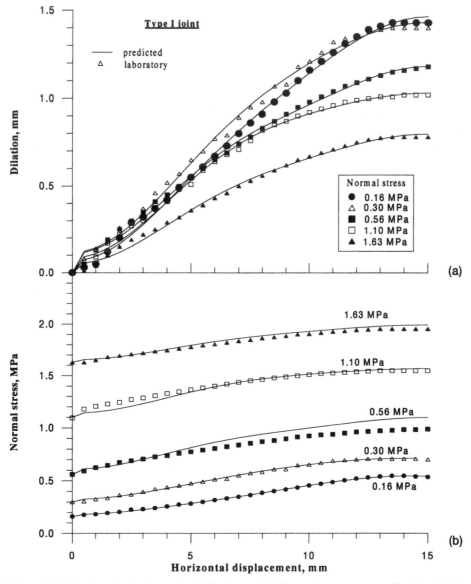

Figure 4.16. Predicted and observed. a) Dilation, and b) Normal stress response for Type I joint.

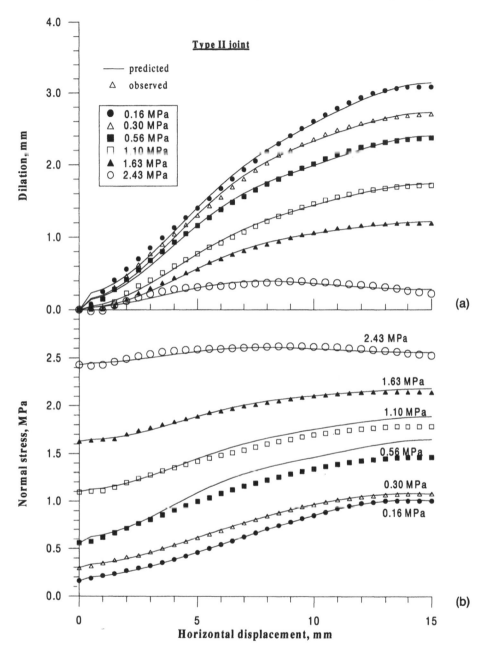

Figure 4.17. Predicted and observed. a) Dilation, and b) Normal stress response for Type II joint.

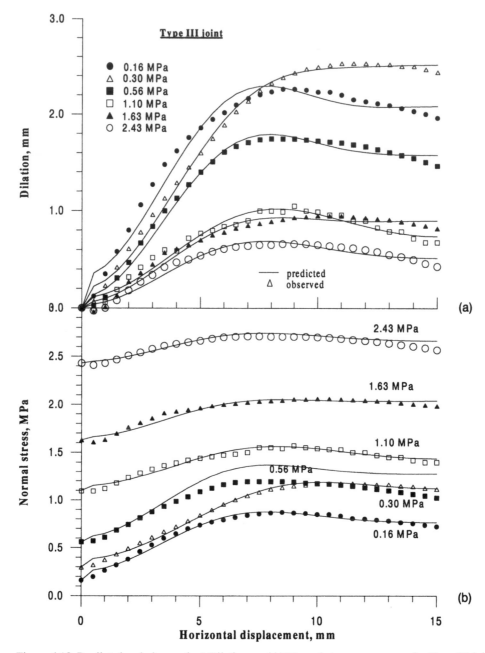

Figure 4.18. Predicted and observed. a) Dilation, and b) Normal stresses response for Type III joint.

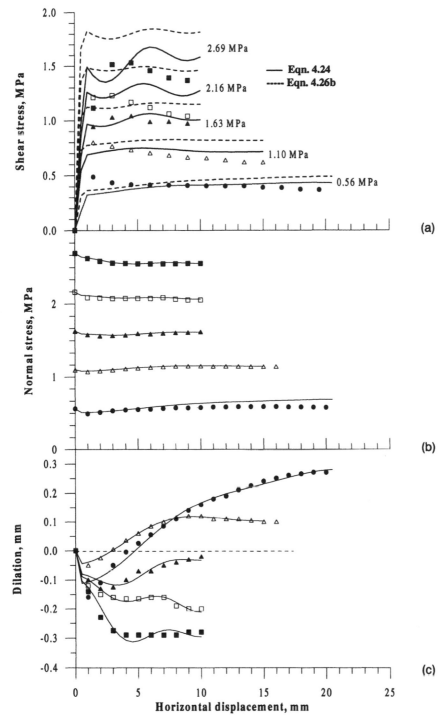

Figure 4.19. Predicted and observed. a) Shear stress, b) Normal stress, and c) Dilation behaviour for natural (field) joint under CNS.

4.7.3 *Shear stress*

Equations 4.24 and 4.26b are used separately to calculate the shear stress with horizontal displacements from the predicted normal stress and dilation behaviour for Type I, II and III prepared joints and natural joints. As expected, the shear behaviour predicted by Equation 4.24 (modified Patton's type) is always smaller than the experimental results, and this discrepancy is larger especially for higher values of σ_{no}, where considerable degradation of asperities occurs (solid lines in Figures 4.19a, 4.20-4.22). However, the predicted shear stress for the smallest asperity angle ($i = 9.5°$) may be considered within reasonable accuracy as the effect of degradation is insignificant. On the other hand, Equation 4.26b (energy balance) can predict the shear stress of saw-tooth joints to match closely with the experimental results, especially in the pre-peak region for any value of σ_{no} (dotted lines in Figures 4.19a, 4.20-4.22). A summary of the predicted peak shear stress (τ_p) and corresponding normal stress (σ_n) for the saw-tooth joints determined by

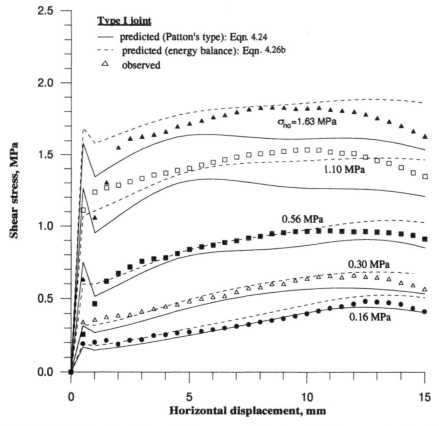

Figure 4.20. Predicted and observed shear stress with horizontal displacement for Type I joint under CNS.

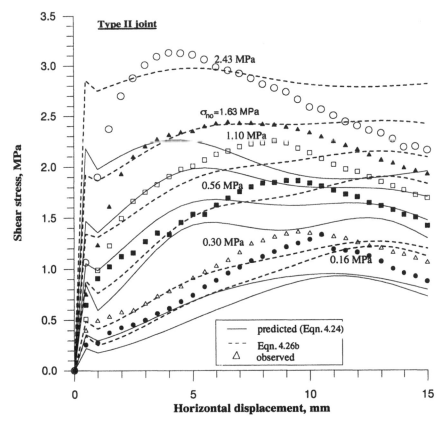

Figure 4.21. Predicted and observed shear stress with horizontal displacement for Type II joint under CNS.

Equations 4.24 and 4.31 is given in Table 4.4. These results confirm that Equation 4.31 is a good choice for shear strength modelling under CNS, if the surface profiles are determined accurately.

As the distribution of asperities for natural joints is not as simple as idealised saw-tooth joints, it becomes very difficult to account for the initial asperity angle (i) in Equations 4.26b and 4.31. The surface profile measurement for these joints indicates that the mean height of the asperities varies in the range of 1.22 to 2.35 mm with various base lengths. Using tilt angle as the initial asperity angle, Equation 4.26b overpredicts the laboratory results. In contrast, Equation 4.24 yielded closer predictions to the laboratory data (Fig. 4.19a). Therefore, unless the surface profiles of rock joints are determined fairly accurately, there is no added benefit in using Equations 4.26b and 4.31 in relation to the predictions based on Equation 4.24.

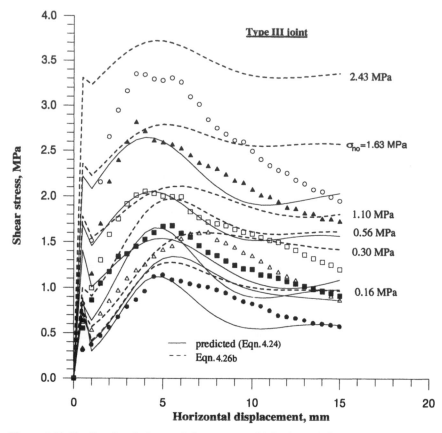

Figure 4.22. Predicted and observed shear stress with horizontal displacement for Type III joint under CNS.

4.7.4 *Strength envelopes*

The computer program automatically executes the peak shear stress and corresponding normal stress for a set of σ_{no}, Fourier coefficients and normal stiffness for modified Patton's model (Eq. 4.24). The strength envelopes for Type I, II and III joints are shown in Figure 4.23 together with the laboratory and proposed shear strength model (Eq. 4.31) predictions. It is observed that for small asperity angles (Type I), the envelopes obtained using Equations 4.24 and 4.31 are very close to the experimental results (Fig. 4.23a). However, for larger asperity angles, the envelope obtained using Equation 4.24 underpredicts the peak friction angle significantly, especially when the normal stress is larger (Figs 4.23b and c). Therefore, considering the overall performance, Equation 4.31 can be used with sufficient accuracy for the prediction of shear strength of joints.

Table 4.4. Laboratory and predicted peak shear stress and corresponding normal stress of unfilled joints.

Joint type	σ_{no}, Mpa	Peak shear stress, Mpa			Corresponding normal stress, Mpa		
		Lab	Eq. 4.31 (new model)	Eq. 4.24 (modified Patton)	Lab	Eq. 4.31 (new model)	Eq. 4.24 (modified Patton)
Type I	0.16	0.48	0.54	0.44	0.53	0.56*	0.53
	0.30	0.66	0.71	0.57	0.70	0.74*	0.69
	0.56	0.97	1.05	0.91	0.94	1.09	1.06
	1.10	1.54	1.52	1.33	1.51	1.57	1.39
	1.63	1.83	1.91	1.64	1.87	1.98	1.81
Type II	0.16	1.33	1.22	0.93	0.88	1.02*	0.92
	0.30	1.36	1.24	0.95	0.94	1.04*	1.01
	0.56	1.86	1.96	1.49	1.29	1.62	1.56
	1.10	2.25	2.20	1.68	1.63	1.85	1.81
	1.63	2.44	2.53	1.99	1.97	2.17	1.92
	2.43	3.12	2.87	2.26	2.57	2.58	2.55
Type III	0.16	1.13	1.21	1.12	0.74	0.91*	0.68
	0.30	1.61	1.73	1.34	1.08	1.25*	0.89
	0.56	1.68	1.84	1.65	1.12	1.39	1.13
	1.10	2.05	1.98	1.66	1.39	1.53	1.31
	1.63	2.82	2.71	2.05	1.91	2.07	1.92
	2.43	3.35	3.50	2.65	2.61	2.71	2.62

* Corrected for normal compliance.

4.7.5 *Infilled joint strength*

After obtaining the peak shear stress of unfilled joints using Equation 4.31, the infilled joint shear strength for various t/a ratios is calculated according to Equation 4.35. The predicted shear strength for various infill thickness to asperity height ratio (t/a) for Type I and II joints is plotted in Figure 4.24 together with the laboratory results. It is observed that the proposed equation (Eq. 4.23) can model the infill joint behaviour very well in relation to the experimental results for a normal stress range of 0.30 to 1.10 MPa. A summary of the predicted shear strengths for Type I and II infilled joints is given in Table 4.5.

4.8 UDEC ANALYSIS OF SHEAR BEHAVIOUR OF JOINTS

4.8.1 *Choice of joint models*

UDEC (Universal Distinct Element Code) has been used successfully to model the shear behaviour of jointed rocks, flow through discontinuities and slope sta-

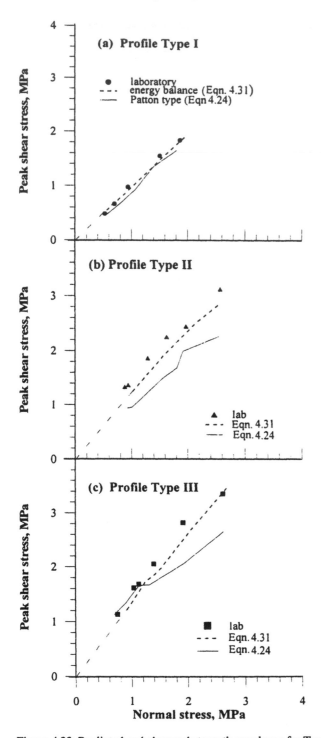

Figure 4.23. Predicted and observed strength envelopes for Type I, II and III joints under CNS.

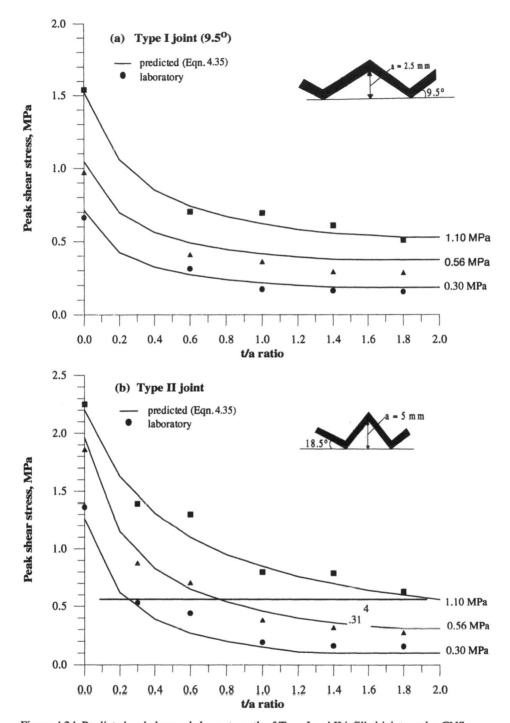

Figure 4.24. Predicted and observed shear strength of Type I and II infilled joints under CNS.

Table 4.5. Experimental and predicted peak shear stress of infilled Type I and II joints.

Joint type	t/a ratio	Laboratory peak shear stress, Mpa			Predicted peak shear stress, MPa (Eq. 4.35)		
		$\sigma_{no} = 0.30$	$\sigma_{no} = 0.56$	$\sigma_{no} = 1.10$	$\sigma_{no} = 0.30$	$\sigma_{no} = 0.56$	$\sigma_{no} = 1.1$
Type I	0	0.66	0.97	1.54	0.71	1.05	1.52
	0.6	0.31	0.41	0.70	0.30	0.50	0.74
	1.0	0.17	0.36	0.69	0.25	0.43	0.62
	1.4	0.16	0.29	0.61	0.22	0.38	0.53
	1.8	0.16	0.29	0.51	0.21	0.38	0.52
Type II	0	1.36	1.86	2.25	1.24	1.96	2.20
	0.3	0.53	0.88	1.39	0.48	1.03	1.45
	0.6	0.44	0.71	1.30	0.27	0.71	1.10
	1.0	0.19	0.38	0.80	0.15	0.52	0.85
	1.4	0.16	0.32	0.79	0.10	0.42	0.70
	1.8	0.16	0.28	0.63	0.10	0.37	0.56

bility problems. The calculations performed in UDEC are based on Newton's second law of motion, conservation of mass and momentum and energy principles. The different joint models that can be used in the UDEC program are outlined below.

The simplest model for simulating discontinuity strength is the linear Mohr-Coulomb friction model. This model is sufficient for smooth discontinuities such as faults at residual strength, which are not non-dilatant. The other models such as the Barton-Bandis model and continuously yielding model may be more appropriate to explain the non-linear behaviour often encountered in rough rock joints. The Barton-Bandis model takes account of more features of discontinuity strength and deformation behaviour than the Coulomb model. However, to apply this model in practical situations, difficulties such as the derivation of relationship for the mobilisation and degradation of surface roughness from a piecewise linear graphical format rather than a formal expression may lead to some irregularities in numerical simulation of the stress-displacement behaviour (Brady & Brown 1993).

The continuously yielding model (Cundall & Hart 1984) is designed to provide a coherent and unified discontinuity deformation and strength model for joints undergoing some elastic and plastic deformations. In the analysis of the shear behaviour of rock joints under constant normal stiffness (CNS) and constant normal load condition (CNL), the continuously yielding joint model is preferred, as it has been shown to have the capability to satisfactorily represent single episodes of shear loading. Further details of the formulation is described in the following section.

4.8.2 *Continuous yielding model*

The continuously yielding joint model (Cundall & Hart 1984) is developed to simulate in a simple manner the internal mechanism of progressive damage of joints under shearing. It is more realistic than the standard Mohr-Coulomb joint model as it considers the non-linear behaviour observed in physical tests such as joint degradation, normal stiffness, dependence on normal stress, and the decrease in dilation angle with plastic shear displacement. The following features make the continuously yielding model more versatile over other models:
- The shear stress versus shear displacement curve approaches a 'target' shear strength of the joint,
- The target shear strength decreases continuously as a function of accumulated plastic displacement which refers to the damage of joints during shearing,
- Dilation angle is taken as the difference between the apparent friction angle and the residual friction angle.

In this model, the normal stress and normal displacement relationship is expressed incrementally as:

$$\Delta\sigma_n = k_n \Delta u_n \tag{4.36}$$

where k_n = normal stiffness = $\alpha_n \sigma_n^{\beta n}$, in which α_n and β_n are model constants.

In shear, the model considers irreversible, non-linear behaviour from the onset of shearing (UDEC manual 1996, page F-2). The displacement relationship is given as:

$$\Delta u_s^p = (1 - F)|\Delta u_s| \tag{4.37}$$

where Δu_s = increment of shear displacement, Δu_s^p = irreversible part of the shear displacement, F = fraction of the current shear stress in relation to the ultimate or limiting shear stress, at the prevailing normal stress.

Thus, the shear strength response is calculated by:

$$\Delta\tau = F k_s \Delta u_s \tag{4.38}$$

where k_s = shear stiffness = $\alpha_s \sigma_n^{\beta s}$ in which α_s and β_s are model constants.

The shear strength is then given by the following equation:

$$\tau_m = \sigma_n \tan\phi_m \, \text{sign}(\Delta u_s) \tag{4.39}$$

where ϕ_m = the friction angle prior to any damage of the asperity.

As damage occurs, this angle is continuously reduced according to the following equation:

$$\Delta\phi_m = -\frac{1}{R}(\phi_m - \phi_b)\Delta u_s^p \tag{4.40}$$

where R = parameter with the dimension of length and related to joint roughness, and ϕ_b = basic friction angle.

The incremental relationship for ϕ_m is equivalent to:

$$\phi_m = \left(\phi_m^i - \phi_b\right) \exp\left(\frac{-u_s^p}{R}\right) + \phi_b \qquad (4.41)$$

where ϕ_m^i = initial friction angle
The effective dilatancy angle is then calculated by:

$$i = \tan^{-1}\left(\frac{|\tau|}{\sigma_n}\right) - \phi_b \qquad (4.42)$$

4.8.3 *Conceptual CNS shear model*

The CNS direct shear boundary conditions used in UDEC are shown in Figure 4.25, which simulate the laboratory conditions as accurately as possible. The material properties of the top block (A) are prescribed in such a way that the external stiffness (8.5 kN/mm) applied by the spring system is modelled explicitly. The blocks B and C are discretised according to the laboratory specimen size, and are assigned the material properties of the model joint tested in this study. The interface asperity shape (triangular) is input by 'crack commands' and subsequently, the joint material properties are assigned by writing subroutines (FISH function).

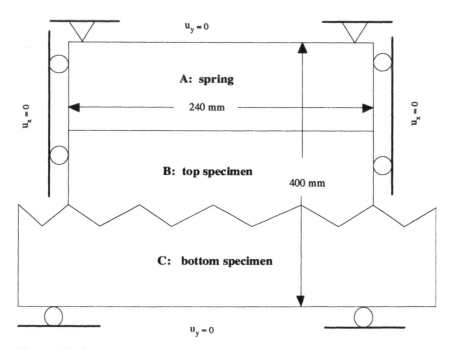

Figure 4.25. Conceptual model for laboratory CNS shear behaviour.

4.8.4 *CNL direct shear model*

The conventional direct shear test (CNL) is modelled in UDEC in a similar way as in CNS, the only exception being that the top block A (Fig. 4.25) is deleted, and the top specimen boundary along the y-direction is made free. This enables dilation to take place under $k_n = 0$ condition. The material and joint properties are kept the same as those under CNS.

4.8.5 *Discretisation of blocks and applied boundary conditions*

Once the blocks are formed and their representative material properties are assigned, they are discretised by a triangular mesh. A low density mesh is used for block A, whereas a fine mesh is prescribed for blocks B and C (Fig. 4.26). In this study, the shear behaviour of Type I joints is modelled under CNS condition.

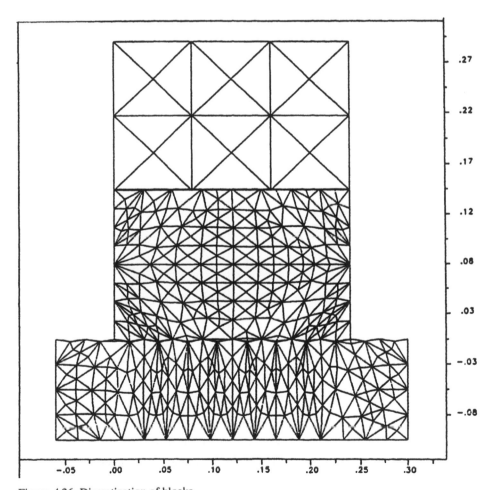

Figure 4.26. Discretization of blocks.

At first, the initial normal stress (σ_{no}) is applied to the joint and the model is allowed to reach equilibrium. A horizontal velocity is applied to the bottom block (C) to produce the required shear displacement compatible with laboratory shear rate. The average normal and shear stresses along the joint are applied using a FISH function. The associated dilation and shear displacements are also calculated via FISH functions. From the shear stress versus displacement plot, the peak shear stress for the applied normal stress can now be determined.

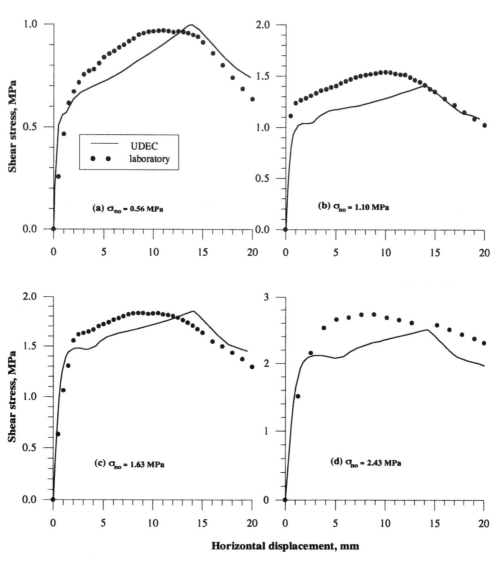

Figure 4.27. UDEC prediction and observed shear stress responses for Type I joints under CNS.

4.8.6 *Results and discussions*

The average shear stress versus horizontal displacement along the joint under CNS is plotted in Figure 4.27 together with the experimental data for selected tests. It is observed that the predicted peak shear stress based on UDEC is in acceptable agreement with the laboratory data, although the pre-peak shear stress response is somewhat underestimated. The predicted shear stress increases with the horizontal displacement and shows a maximum value approaching the peak-to-peak contact of the asperities. In contrast, the laboratory peak shear stress is observed to occur at a smaller horizontal displacement, in comparison with UDEC analysis. At high levels of initial normal stress (σ_{no}), the asperity crushing is significant as reflected by much smaller dilation than that shown by UDEC data (Fig. 4.28d), and in this particular case, UDEC prediction of shear stress is considerably smaller than the observed data (Fig. 4.27d). Asperity degradation cannot be modelled using UDEC, which tends to under-predict the shear stress values. This can be clearly observed in Figure 4.28, where a pure frictional behaviour with enhanced dilation is shown by UDEC analysis. It is not surprising to note that the predicted dilation versus shear displacement curves (Fig. 4.28) follow the same shape of the asperities, in the absence of any asperity breakage.

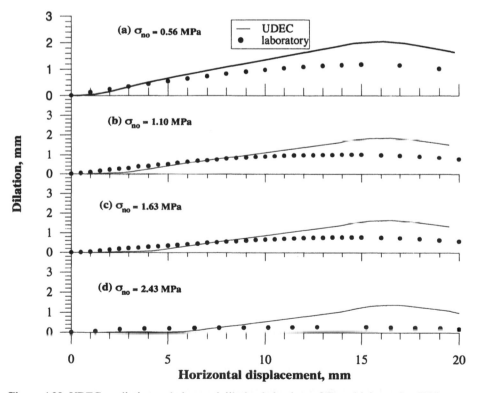

Figure 4.28. UDEC prediction and observed dilation behaviour of Type I joint under CNS.

Under CNS, the normal stress increases during asperity overriding, and if the current normal stress (σ_n) exceeds the compressive strength of asperities, degradation occurs. The maximum possible normal stress should occur if the peak to peak contact of asperities would take place. However, if the initial normal stress, σ_{no} is high, degradation occurs even before the peak to peak contact of asperities is reached. Under CNL, the laboratory observations verified that the asperity degradation was less prominent at the same initial normal stress and at similar shear displacements.

The average normal stress under CNS along the joints for various σ_{no} values is plotted in Figure 4.29. As the dilation of the joint is overestimated by UDEC, the corresponding normal stress is also overpredicted. The peak shear stress occurs at a greater shear displacement for UDEC in comparison with the laboratory results. After attaining the peak, the shear stress continues to decrease with the shear displacement of the joint. The predicted shear stress versus horizontal displacement plot under CNL is shown in Figure 4.30 for comparison. It is observed that under CNL, the peak shear stress is attained at a smaller horizontal displacement in comparison with CNS. However, UDEC predictions are closer to the observed laboratory data, as the asperity crushing is less significant under CNL. This leads to the conclusion that UDEC is more appropriate in modelling CNL behaviour at insignificant asperity degradation. It is beyond the scope of this study to attempt to modify UDEC capabilities to accommodate realistic asperity crushing, and thereby accurately model shearing under the CNS condition.

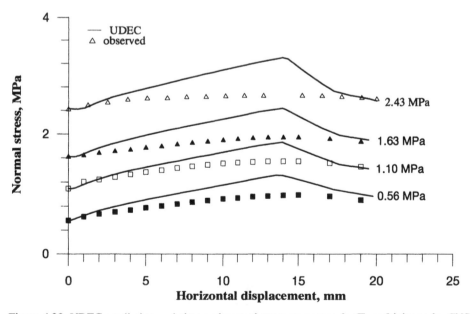

Figure 4.29. UDEC prediction and observed normal stress responses for Type I joint under CNS.

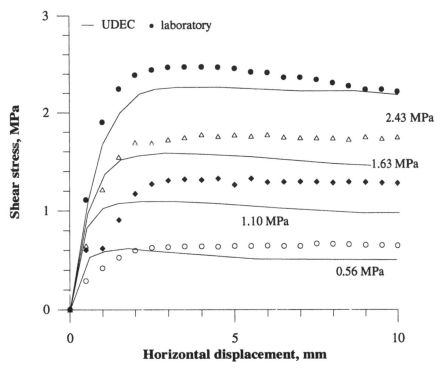

Figure 4.30. UDEC prediction and laboratory shear behaviour of Type I joint under CNL.

4.9 SUMMARY OF SHEAR STRENGTH MODELLING

A new shear strength model for predicting the behaviour of unfilled and infilled joints under constant normal stiffness condition (CNS) was formulated based on the Fourier Transform method, energy balance principle and the hyperbolic stress-strain representation. A conceptual model was also presented to analyse the shear behaviour of unfilled joint using the Universal Distinct Element Code (UDEC) in a simplified manner. The performance of the new shear strength model based on Equations 4.26b and 4.31 is discussed in detail, and compared with other available methods such as the modified Patton (1966) and Barton (1973) models to some extent. Important aspects of the shear strength models are summarised below:

– The proposed model described in this thesis (Eq. 4.26b) is a significant extension of Patton (1966), to successfully model the shear behaviour of soft joints under CNS, especially with regard to the variations in dilation, and the shear and normal stress response with horizontal displacement.
– The peak shear stress model incorporating the Fourier Coefficients (Eq. 4.31) are proposed to predict the shear strength of joints under CNS with acceptable accuracy, in relation to laboratory data.

– The infill joint behaviour can be modelled accurately using the new shear
  strength criterion for infilled joint as represented by Equation 4.35.
– If UDEC is employed to predict the peak shear stress of unfilled joints under
  CNS, it overestimates the joint dilation as well as the normal stress. The maximum peak shear stress in UDEC is attained at a greater shear displacement in
  comparison with the laboratory data. In contrast, the UDEC predictions are
  generally in good agreement with laboratory data under CNL condition, where
  the asperity degradation was found to be less significant.

REFERENCES

Barla, G., Forlati, F. & Zaninetti, A. 1985. Shear behaviour of filled discontinuities. *Proc. Int. Symp.
on Fundamentals of Rock Joints*, Bjorkliden, pp. 163-172.

Barton, N. 1973. Review of a new shear strength criterion for rock joints. *Engineering geology*,
7: 287-332.

Barton, N., Bandis, S. & Bakhtar, K. 1985. Strength, deformation and conductivity coupling of rock
joints. *Int. J. Rock. Mech. Min. Sci. & Geomech. Abstr.*, 22(3): 121-140.

Brady, B.H.G. & Brown, E.T. 1993. *Rock mechanics for underground mining* (2nd edition). Chapman & Hall, 571 p.

Cundall, P.A. & Hart, R.D. 1984. *Analysis of block test No. 1 – Inelastic Rock Mass Behaviour:
Phase 3 – A characterisation of joint behaviour (Final Report)*. Itasca Consulting Group Report,
Rockwell Handford Operations, Subcontact SA-957 (cited from UDEC manual).

de Toledo, P.E.C. & de Freitas, M.H. 1993. Laboratory testing and parameters controlling the shear
strength of filled rock joints. *Geotechnique*, 43(1): 1-19.

Duncan, J.M. & Chang, C.Y. 1970. Nonlinear analysis of stress and strain in soils. *J. Soil Mech. &
Foundation Div.*, ASCE, 96(5): 1629-1653.

Haberfield, C.M. & Johnston, I.W. 1994. A mechanistically based model for rough rock joints. *Int.
J. Rock. Mech. Min. Sci. & Geomech. Abstr.*, 31(4): 279-292.

Haberfield, C.M. & Seidel, J.P. 1998. Some recent advances in the modelling of soft rock joints in
direct shear. *Proc. Int. Conf. Geomech./Ground Control in Mining & Underground Construction*,
Wollongong, Aziz, N. & Indraratna, B. (eds), 1: 71-84.

Haque, A. 1999. Shear behaviour of rock joints under constant normal stiffness. Ph.D thesis, University of Wollongong, 271 p.

Heuze, F.E. & Barbour, T.G. 1982. New models for rock joints and interfaces. *J. Geotech. Eng.
Div., ASCE*, 108(GT5): 757-776.

Indraratna, B. 1990. Development and applications of a synthetic material to simulate soft sedimentry rocks. *Geotechnique*, 40(2): 189-200.

Indraratna, B., Haque, A. & Aziz, N. 1999. Shear behaviour of idealised infilled joints under constant normal stiffness. *Geotechnique*, 49(3): 331-355.

Indraratna, B., Herath, A. & Aziz, N. 1995. Characterisation of surface roughness and its implica-
tions on the shear behaviour of joints. *Mechanics of Jointed and Faulted Rocks*, Rossmanith
(ed.), Balkema Publishers, Rotterdam, pp. 515-520.

Johnston, I.W. & Lam, T.S.K. 1989. Shear behaviour of regular triangular concrete/rock joints-
analysis. *J. Geotech. Eng., ASCE*, 115(5): 711-727.

Johnston, I.W., Lam, T.S.K. & Williams, A.F. 1987. Constant normal stiffness direct shear testing
for socketed pile design in weak rock. *Geotechnique*, 37(1): 83-89.

Kodikara, J.K. & Johnston, I.W. 1994. Shear behaviour of irregular triangular rock-concrete joints. *Int. J. Rock Mech. Min. Sci. & Geomech. Abstr.*, 31(4): 313-322.

Ladanyi, B. & Archambault, G. 1970. Simulation of shear behaviour of a jointed rock mass. *Proc. 11th Symp. on Rock Mechanics*, Urbana, Illinois, pp. 105-125.

Ladanyi, H.K. & Archambault, G. 1977. Shear strength and deformability of filled indented joints. *Proc. 1st Int. Symp. Geotech. Structural Complex Formations*, Capri, pp. 317-326.

Maksimovic, M. 1996. The shear strength components of a rough rock joint. *Int. J. Rock. Mech. Min. Sci. & Geomech. Abstr.*, 33(8): 769-783.

Milovic, D.M., Trizot, G. & Trumier, J.P. 1970. Stresses and displacements in an elastic layer due to inclined and ecentric load over a rigid strip. *Geotechnique*, 20: 231-252.

Newland, P.L. & Alley, B.H. 1957. Volume changes in drained triaxial tests on granular materials. *Geotechnique*, 7: 17-34.

Papaliangas, T., Hencher, S.R., Lumsden, A.C. & Manolopoulou, S. 1993. The effect of frictional fill thickness on the shear strength of rock discontinuities. *Int. J. Rock Mech. Min. Sci. & Geomech. Abstr.*, 30(2): 81-91.

Patton, F.D. 1966. Multiple modes of shear failure in rocks. *Proc. 1st Cong. ISRM*, Lisbon, pp. 509-513.

Phien-wej, N., Shrestha, U.B. & Rantucci, G. 1990. Effect of infill thickness on shear behaviour of rock joints. *Rock Joints*, Barton & Stephansson (eds), Balkema Publisher, Rotterdam, pp. 289-294.

Saeb, S. & Amadei, B. 1990. Modelling joint response under constant or variable normal stiffness boundary conditions (Technical note). *Int. J. Rock Mech. Min. Sci. & Geomech. Abstr.*, 27(3): 213-217.

Saeb, S. & Amadei, B. 1992. Modelling rock joints under shear and normal loading. *Int. J. Rock Mech. Min. Sci. & Geomech. Abstr.*, 29(3): 267-278.

Seidel, J.P. & Haberfield, C.M. 1995. The application of energy principles to the determination of the sliding resistence of rock joints. *Rock Mech. & Rock Eng.*, 28(4): 211-226.

Skinas, C.A., Bandis, S.C. & Demiris, C.A. 1990. Experimental investigations and modelling of rock joint behaviour under constant stiffness. *Rock Joints*, Barton & Stephansson (eds), Balkema Publisher, Rotterdam, pp. 301-307.

Spiegel, M.R. 1974. *Theory and problems of Fourier analysis with application to boundary value problems*. Schaum's outline series, McGraw-Hill Inc., 191 p.

UDEC users manual version 3.0. 1996. Itasca Consulting Group, Inc., Minneapolis, USA.

CHAPTER 5

Simplified approach for using CNS technique in practice

5.1 INTRODUCTION

This chapter deals with the application of CNS technique in practical problems (e.g. underground excavation in jointed rocks, slope stability analysis). In the analysis, the contribution of the surrounding rock mass stiffness is considered to be constant, where the normal stress continues to vary during deformation. The laboratory results presented in Chapter 3 reveal that the CNL method always overestimates the peak friction angle, hence, overpredicts the joint shear strength. Therefore, predictions made on the basis of CNL condition may not always be representative of the actual field conditions, where the normal load does not remain stable. As pointed out in Chapter 2, CNS technique is more appropriate for the stability analysis of the roof of an excavation, rock socketed piles, bolted joints etc. In this study, the stability analysis of a typical underground roadway in jointed rock using the Universal Distinct Element Code (UDEC) is presented. The safety analysis of a slope supported by grouted bolts is also illustrated in a simplified manner.

5.2 UNDERGROUND ROADWAY IN JOINTED ROCK

Figure 5.1 shows a typical underground roadway excavated in a horizontally bedded rock mass. As the roadway is excavated, the convergence of the roof will be controlled by the stiffness of the bedded strata (assumed to be constant), while σ_n acting perpendicular to the bedded planes will vary during and after excavation. Therefore, using CNS strength parameters as input data in UDEC should produce more realistic results. Applicability of the CNL strength parameters in this situation could overestimate stability. The results of the analysis of a roadway excavation are discussed below.

5.2.1 *Boundary conditions*

A compressive stress of 5 MPa is applied in the x-direction and a 10 MPa compressive stress is applied in the y-direction (top). The bottom boundary is fixed

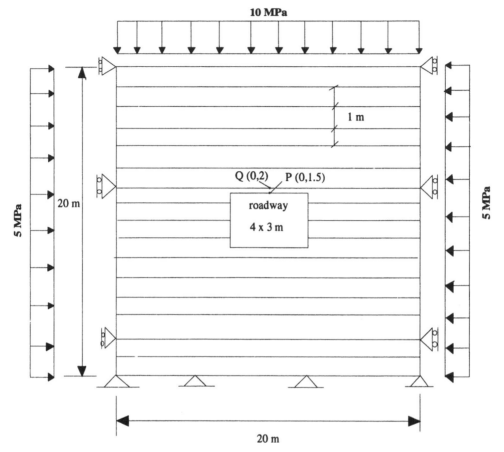

Figure 5.1. Modelling of a roadway in horizontally bedded strata with boundary conditions.

from moving in the *y*-direction as shown in Figure 5.1. FISH functions are used to formulate the model and assigning material properties. The model is then allowed to reach the equilibrium condition under the applied boundary loads.

5.2.2 *Roadway excavation*

Once the model has reached an equilibrium state, the roadway is excavated. Figure 5.2 shows the variations in displacement (u_y) at the top of roadway P (0,1.5) and Q (0,2) with time. It is observed that the roof has attained equilibrium after approximately 60 mm of displacement along the *y*-direction. The ground reaction curve (GRC) shown in Figure 5.3 reveals that the normal stress (σ_y) at point P (0,1.5) decreases linearly upto a displacement of 20 mm implying that constant normal stiffness is maintained. After this, the variation of normal stress with displacement is marginal, and ultimately, a compressive stress of approximately 0.30 MPa is reached, beyond which no change in σ_y was observed. In accordance, the

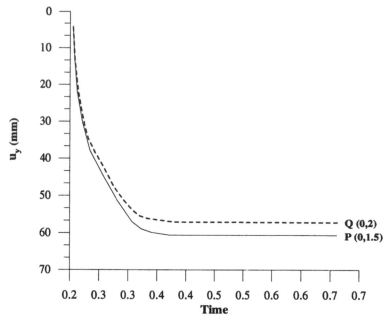

Figure 5.2. Vertical displacements at points *P* (0,1.5) and *Q* (0,2) on the roof of roadway.

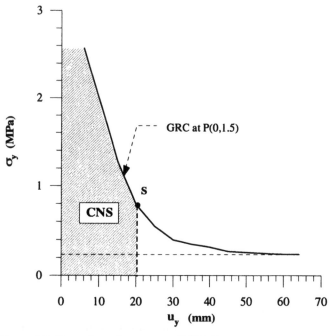

Figure 5.3. Ground reaction curve at *P* (0,1.5).

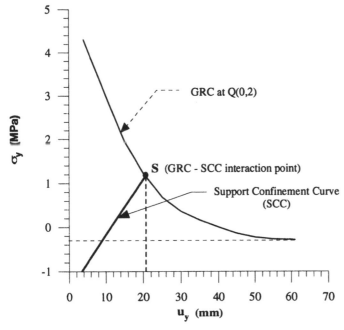

Figure 5.4. Ground reaction curve at Q (0,2).

GRC shows that the roadway roof has attained stability. Similar behaviour is also observed for Q (0,2) point on the roof of the roadway (Fig. 5.4).

The ground reaction curve determined from UDEC analysis represents a typical stable excavation. It also shows the variation of the normal stress with displacement. Figures 5.3 and 5.4 reveal that the σ_y variation is linear for a substantial displacement of the roof. If such an excavation requires support to prevent failure, then bolts and/or other supports may be installed to interact at point S on the GRC (Fig. 5.4). This will minimise the support load and would not result in unsafe roof deformation. The CNS parameters obtained from laboratory testing will be more appropriate for support design, as the GRC can be employed successfully for the stress-deformation domain compatible with the constant normal stiffness. The constant normal load condition will rarely be encountered above a mine excavation, where σ_n continuously varies with displacement, until such time an equilibrium load is established after support installation. Such equilibrium, however, often takes a significant period of time.

5.3 STABILITY ANALYSIS OF SLOPE

Figure 5.5 shows a rock slope, where a major discontinuity plane passes through the inclined face of the slope at an angle β. The discontinuity is assumed to have a very rough surface. The slope is supported by untensioned fully grouted bolts. The passive supports would work effectively if the discontinuity plane dilates during shear movement. This dilation would generate a tensile force in the bolt, its magnitude depending on the bolt-grout stiffness (*EA/L*), thereby, producing an additional normal load on the discontinuity plane (Fig. 5.6). The normal load acting perpendicular to the plane will not reach a constant value during joint displacement, but the stiffness of the bolt-grout-rock composite will remain constant at small to moderate strains which do not initiate bolt or grout yielding. Therefore, CNS rather than CNL can explain the shear behaviour of such a discontinuity plane better. A simplified, comparative stability analysis of this slope is given in the subsequent section. The illustrated example of the slope is based on the di-

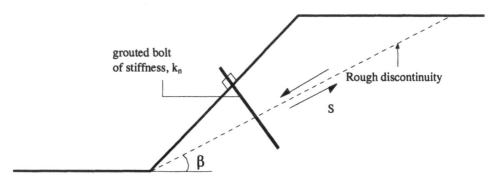

Figure 5.5. A slope supported by untensioned grouted bolt.

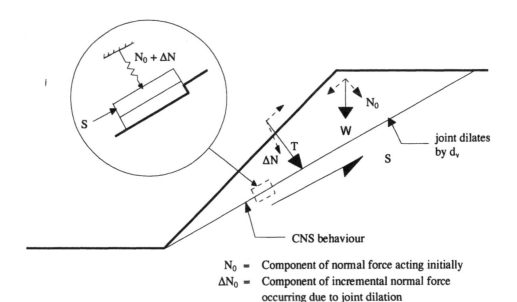

N_0 = Component of normal force acting initially
ΔN_0 = Component of incremental normal force occurring due to joint dilation
S = shear resistance

Figure 5.6. Forces acting on the discontinuity plane after displacement.

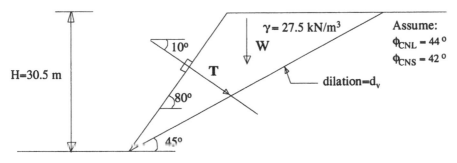

Figure 5.7. Simplified rock wedge indicating geometry and rock propereties.

mensions and properties shown in Figure 5.7, for a potentially unstable wedge at the Kangaroo Valley site, in NSW, Australia.

5.3.1 *Limit equilibrium analysis (initial condition without bolts)*

The weight of the block

$$
\begin{aligned}
W &= 0.5 \, \gamma \, H^2 \, (\cot \alpha - \cot \beta) \\
&= 0.5 \times 27.5 \times (30.5)^2 \times (\cot 45° - \cot 80°) \\
&= 10.54 \times 10^3 \text{ kN/m length of slope}
\end{aligned}
$$

Resolving the force, W, into components parallel and perpendicular to the discontinuity plane, the disturbing force (DF) and resisting force (RF) can be calculated as:

$$
\begin{aligned}
DF &= W \sin 45° \\
&= 10.54 \times 10^3 \times \sin 45° \\
&= 7.45 \times 10^3 \text{ kN/m}
\end{aligned}
$$

The resisting force (RF) can be calculated using $\phi_{CNL} = 44°$ (determined from laboratory CNL test), which is typical of sandstone joints obtained at Kangaroo Valley.

$$
\begin{aligned}
RF &= (W \cos 45°) \times \tan \phi_{CNL} \\
&= 10.54 \times 10^3 \times \cos 45° \times \tan 44° \\
&= 7.19 \times 10^3 \text{ kN/m}
\end{aligned}
$$

Factor of safety (FS)

$$
\begin{aligned}
FS &= RF / DF = (7.19 \times 10^3) / (7.45 \times 10^3) = 0.96 \\
&= 0.96 < 1.0 \text{ indicates instability}
\end{aligned}
$$

The above factor of safety suggests that rock bolts are needed to stabilise the joint.

5.3.2   *CNS analysis (i.e. considering bolt contribution)*

The unstable joint is supported by say ten fully grouted untensioned bolts of diameter 2.5 cm. The factor of safety (*FS*) of this slope is then calculated based on the following simplified procedure. As the joint dilates by δ_v, the bolt is tensioned and the associated increase in normal load on the discontinuity plane becomes beneficial. Now it is required to calculate the bolt tension developed due to the dilation, δ_v for the given rough joint surface. Two assumptions are made for the simplicity of calculations:

1. All the bolts contribute to an equal load, and
2. Uniform (equivalent) stress variation along the bolt length.

The tension force (*T*) developed in the grouted bolt considering an effective bolt diameter of 2.5 cm and an effective grouted length of $L = 1.0$ m is given by:

$$T = EA\delta_v/L$$

(for convenience, the stiffness of the grout annulus is neglected)

where E = Youngs modulus (for steel = 200×10^6 kPa), and δ_v = joint dilation (mm).

$$T = \frac{200 \times 10^6 \times \frac{\pi}{4}\left(2.5 \times 10^{-2}\right)^2 \times \delta_v \times 10^{-3}}{1} = (98 \times \delta_v) \text{ kN per bolt}$$

By knowing the Fourier coefficients of this particular joint, the dilation behaviour of the joint system can be predicted at any normal stress using the methodology as described in Chapter 4. Once the dilation characteristics are known, normal stress and shear stress responses of the joint can be modelled accurately from the proposed shear strength model.

After a dilation of δ_v, the total normal force acting on the plane becomes;

$$N = N_0 + \Delta N$$

which can be obtained by adding up the normal components of forces *W* and *T*, as explained below.

$$\begin{aligned}
N &= W \cos 45° + nT \sin (45° + 10°) \\
&= 10.54 \times 10^3 \cos 45° + 10 \times (98 \times \delta_v) \sin 55°
\end{aligned}$$

Assuming that the joint is frictional and using the CNS test angle, $\phi_{CNS} = 42°$ ($\phi_{CNS} < \phi_{CNL}$), the resisting force (*RF*) can then be calculated as:

$$\begin{aligned}
RF &= n (98 \times \delta_v) \cos 55° + N \times \tan \phi_{CNS} \\
&= 10 \times (98 \times \delta_v) \cos 55° + \\
&\quad [10.54 \times 10^3 \cos 45° + 10 \times (98 \times \delta_v) \sin 55°] \tan 42°
\end{aligned}$$

The disturbing force (*DF*) acting along the plane is given by:

$$DF = W \times \sin 45° = 7.45 \times 10^3 \text{ kN/m}$$

In this manner, the *FS* of the slope against stability is calculated for various magnitudes of dilation of joints using CNS shear strength parameters. The corresponding values of *FS* are summarised in Table 5.1. As expected, the greater the joint dilation (i.e. the degree of roughness is higher), the greater the factor of safety. The values of dilation are assumed for demonstration purpose of shearing mechanism involved in an arbitrary joint whose surface profile characteristics are unknown, hence the Fourier transform method cannot be used.

5.3.3 *CNL analysis considering joint contribution*

In this analysis, the normal load acting on the discontinuity plane for any dilation, δ_v is considered, and the resisting force (*RF*) is calculated based on the friction angle ϕ_{CNL}. The previous calculation (Section 5.3.1) is revised accordingly, and the resisting and disturbing forces are determined below.

According to this analysis,

$$RF = n \,(98 \times \delta_v)\cos 75° + N \times \tan \phi_{CNL}$$
$$= 10 \times (98 \times \delta_v)\cos 55° +$$
$$[10.54 \times 10^3 \cos 45° + 10 \times (98 \times \delta_v)\sin 55°]\tan 44°$$

and

$$DF = W \times \sin 45° = 10.54 \times 10^3 \sin 45° = 7.45 \times 10^3 \text{ kN/m}$$

The corresponding factor of safety is given in Table 5.2, for varying dilation.

Table 5.1. Factor of safety determined on the basis of CNS.

Dilation, δ_v (mm)	Resisting force (*RF*) (kN)	Disturbing force (*DF*) (kN)	*FS = RF/DF*
1	7992	7450	1.07
2	9278	7450	1.24
3	10563	7450	1.42
4	11848	7450	1.59

Table 5.2. Factor of safety determined on the basis of CNL.

Dilation, δ_v (mm)	Resisting force (*RF*) (kN)	Disturbing force (*DF*) (kN)	*FS = RF/DF*
1	8531	7450	1.13
2	9869	7450	1.32
3	11206	3818	1.50
4	12544	3818	1.68

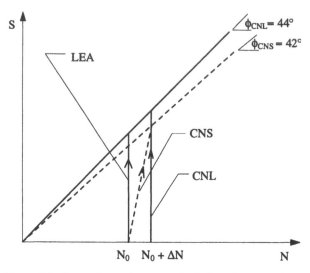

Figure 5.8. Comparison of stress-paths followed in the safety analysis.

It is clear from Tables 5.1 and 5.2 that the *FS* determined under CNL is somewhat greater than that of CNS for the same problem. Therefore, from a stability point of view, the lower *FS* (i.e. CNS values) should be considered as more critical in design and stability assessment.

In summary, the analysis of the *FS* under CNS and CNL can be explained as follows:
– The initial behaviour is represented by Limit Equilibrium Analysis (LEA) path (Fig. 5.8), where the role of deformation during slope movement is not included in the computations (i.e. only stress/forces are considered).
– Once the bolts are installed, the dilation is controlled by the bolt-grout stiffness, and an increase in normal stress occurs due to bolt tension. Therefore, the behaviour is closer to the CNS path (Fig. 5.8).
– If the maximum value of the normal force (i.e. $N_0 + \Delta N$) is used to analyse the slope based on the CNL approach, then the CNL path (Fig. 5.8) is followed, which clearly shows that the normal load is assumed to be constant throughout shearing. In real situations, if the normal stress is assumed to be constant, the CNL approach overestimates the *FS* in comparison to CNS (Tables 5.1 and 5.2).

CHAPTER 6

Highlights of rock joint behaviour under CNL and CNS conditions, and recommendations for the future

6.1 SUMMARY

The shear strength characteristics of unfilled and infilled soft joints were described by the authors with reference to Constant Normal Stiffness (CNS) and Constant Normal Load (CNL) conditions. The laboratory investigations described:
– CNS direct shear test under various shear displacement rate,
– CNL and CNS tests on unfilled saw-tooth and natural (tension) joints, and
– CNS tests on infilled joints.
In this book, the behaviour of synthetic joints based on gypsum plaster, cement-sand mixture etc. were discussed, while some studies using natural joints were also described.

Following Chapter 1 which introduced the aims and scope of this book in relation to the current status on the understanding of the shear behaviour of soft unfilled and infilled rock joints, Chapter 2 was devoted to a critical review of the behaviour of unfilled joints under CNS condition with a special focus on soft joints. The interpretation of data relates to the stress paths, stress-strain relationship, strength envelope and the dilation behaviour of the joints. The principles and the design features of the large-scale CNS direct shear apparatus used by the authors at University of Wollongong, Australia were discussed. The analytical models associated with unfilled joint behaviour under CNS were also highlighted.

The various influential factors governing the shear behaviour of infilled joints and the models available for the simulation of infill joint behaviour were discussed in Chapter 3. As there have been very limited studies on infilled joints tested under CNS, both CNL and CNS test procedures and results were discussed.

Chapter 4 contained the available shear strength models and a new shear strength model developed by the authors which is applicable to both infilled and unfilled soft joints under CNS condition. A comparative study of the authors' model predictions with laboratory results as well as the predictions of modified Patton (1966) and Barton (1973) models were presented graphically and in tabular form. The modelling aspects of unfilled joint behaviour using UDEC were also presented.

Chapter 5 presented a simplified analysis and a set of guidelines with regard to practical application of the CNS testing methodology. The potential use of the new shear strength model to a slope stability problem and to an underground excavation (e.g. mine roadway) was also highlighted.

The general conclusions are presented below, in addition to the specific concluding comments made under individual chapters.

6.1.1 *Behaviour of unfilled joints under various rates of shear displacement*

The increase in the rate of shear displacement contributes to the increase in the shear strength of soft unfilled joints under CNS condition.

The rate of shear displacement below 0.5 mm/min. has very little effect on the shear strength of joints.

6.1.2 *Behaviour of unfilled joints under CNL and CNS*

The shear behaviour of rock joints under Constant Normal Stiffness (CNS) is closer to reality than the Constant Normal Load (CNL) testing for certain field applications, particularly in mining excavations (e.g. above a mine roadway).

The CNS shear strength envelope always plots below that of the CNL.

The horizontal displacement corresponding to peak shear stress under CNL is always smaller than that of CNS for a given initial normal stress (σ_{no}). The horizontal displacement required to attain the peak shear stress decreases with the increase in σ_{no} for both CNL and CNS conditions.

Compared to the CNS method, the CNL method overestimates the joint dilation, hence, underestimates the joint normal stress during shearing.

The CNL strength envelope always yields a higher friction angle in comparison with the CNS strength envelope.

6.1.3 *Shear behaviour of unfilled and natural joints under CNS*

The shear strength of a joint increases with the increase in asperity angle under a given initial normal stress (σ_{no}), but to a different extent in comparison with the CNL condition.

The asperity degradation under high initial normal stress is significant, indicating a shear behaviour similar to that of the CNL condition.

The strength envelope is almost linear for smaller asperity angles and nonlinear for larger asperity angles or increased surface roughness under CNS condition.

The CNS method enables the strength envelope to be drawn from a few selected tests on joints, whereas a relatively large number of tests are needed to obtain the CNL envelope. This is because the stress paths tend to travel along the CNS strength envelope, especially at smaller σ_{no} values. At larger σ_{no} values, the stress paths just reach the strength envelope (but do not travel along it), and rapidly drop in shear stress thereafter.

6.1.4 *Behaviour of Infilled joints under CNS condition*

The shear strength of unfilled joints decreases rapidly with the addition of a very thin layer of infill.

The shear displacement corresponding to the peak shear stress increases with the increase in infill thickness, until the infill thickness exceeds the height of the asperity.

The shear behaviour of infilled joints becomes similar to that of the infill behaviour itself, when the infill thickness to asperity height (t/a) ratio exceeds 1.40. This t/a ratio was considered as the 'critical' ratio for CNS condition in this study. For CNS tests conducted on the soft simulated joints, the 'critical' ratio varies from about 1.4 to 1.8, as the applied initial normal stress is increased.

The joint behaviour is dilatant below the critical t/a ratio, and fully compressive above it.

Laboratory observations indicate that the shear plane passes through the asperity and infill for a t/a ratio less than unity, and touches the 'crown' of the toothed asperity for t/a ratios close to one. The shear plane passes only through the infill for t/a ratios greater than the critical value, depending upon the initial normal stress, σ_{no}.

The stress path plots clearly reveal the role of the critical t/a ratio. The stress paths plot to the left once the critical t/a ratio is exceeded (i.e. continual reduction in normal stress and rapid increase in shear stress under very small displacement, followed by a gradual decrease). If the critical t/a ratio is not exceeded, the stress-paths plot to the right, indicating an increase in normal stress until the shearing of all the asperities is completed.

The peak friction angle of clean joints considerably decreases after the addition of a thin infill layer of bentonite. In particular, the decrease in friction angle with the infill thickness is more pronounced as the asperity angle is increased.

6.1.5 *New shear strength model by the authors*

The Fourier transform method was introduced to model the dilatancy behaviour of unfilled joints under the CNS condition. It was verified that the proposed method could accurately predict the dilation of joints corresponding to a given shear displacement in relation to the measured data. For a given initial normal stress (σ_{no}), the predicted dilation can be used to determine the shear stress-normal stress relationships for a specific joint, based on the proposed shear stress equation (Eq. 4.26b) using a computer subroutine.

The proposed shear strength model (Eq. 4.31) can predict the laboratory behaviour with acceptable accuracy, and can be used in confidence for field situations if the surface roughness of the natural joints can be estimated.

The drop in shear strength with the increasing t/a ratio can be fitted to a hyperbolic decay model. Knowing the shear strength of the unfilled joints, the shear strength of the infilled joints can be estimated for a given t/a ratio and σ_{no} value using the shear strength relationship (Eq. 4.35) developed for infilled joints.

The Universal Distinct Element Code (UDEC) can model the laboratory shear behaviour of unfilled joints with acceptable accuracy under low applied normal stress, where asperity crushing is insignificant. Under high normal stress levels, the predicted shear stress is much smaller than the laboratory values, because asperity crushing becomes significant. The predicted dilation is overestimated especially when the shearing of asperities takes place, hence, the corresponding normal stress is also overpredicted.

6.2   RECOMMENDATIONS FOR FURTHER STUDY

6.2.1   *Modifications to laboratory procedures*

The shear behaviour of idealised infilled joints evaluated under CNS by the authors can be extended to characterise and model the behaviour of natural joints, in the following manner:
– Surface profile of the natural joint specimens should also be mapped accurately using the Coordinate Measuring Machine (CMM), prior to testing,
– Infill material collected from the field should be tested at varying thickness to determine the appropriate 'critical' ratio which is truly representative of the field conditions,
– All jointed specimens should be tested under CNS, where the applied initial normal stress (σ_{no}) would simulate the field conditions.

6.2.2   *Field mapping*

The shear strength model proposed by the authors can be extended to predict the field shear strength of joints once the surface geometry is accurately mapped. In order to achieve this, either Fractal method or Fouier Transformation could be used. Further validation of the authors' model based on more extensive field data obtained from various mine sites is desirable in order to apply the model for practical geotechnical problems.

6.2.3   *Effective stress approach*

The majority of current rock joint models are capable of predicting the shear behaviour of relatively dry joints for simplified joint surfaces. Many of these models do not include the complex joint surface characteristics, the effects of infill properties, and the degradation behaviour of asperities. In the field, joints are often infilled with clay and silt, which can be fully or partially saturated. Accordingly, the correct shear behaviour is very different from the simplified behaviour predicted for dry joints. A realistic and comprehensive shear strength model should be based on a thorough effective stress analysis, considering the role of internal water pressures, where warranted. In addition, it should also address the role of Constant Normal Stiffness (CNS) boundary condition, which is commonly encountered in the field, for instance in underground mining situations. In order to achieve this, a CNS apparatus would require some modification to investigate the effect of pore pressure development during shearing on the effective shear strength of the rock joints.

6.2.4 *Bolted joints*

The CNS technique could be extended to model the behaviour of bolted joints, in view of the stabilisation of unstable roofs in underground excavations. In the field, the shearing of bolted interface and the behaviour of bolt-rock interface may take place under CNS condition. Therefore, to find the true bolt capacity, it is important to know the dilation characteristics of the bolt surface-rock interface under the CNS condition.

An investigation of this problem is now underway at University of Wollongong, Australia, in collaboration with underground mining industry.

6.2.5 *Scale effects*

Rock masses are inhomogeneous and form discontinuous media. Therefore, the results obtained from laboratory testing mostly depend on the sample size and the location from where the specimens are collected. The variation of properties with the size of the specimens is known as the Scale Effect. For intact rocks, the increase in specimen size decreases the strength of the rock (Barton 1987, Hoek & Brown 1980, Yoshinaka & Yamabe 1981, Herget 1988, Bandis 1990). However, the scale effect has insignificant influence on the shear strength of infilled rock joints (Salas 1968, Londe 1973, Infanti & Kanji 1978). This is due to the ratio of infill thickness to asperity height, which governs the shear behaviour of infilled joints. The change of joint size may affect the shear strength, peak shear displacement, shear stiffness, and peak dilation angle of non-planar joints (Barton & Choubey 1977, Leichnitz & Natau 1979). As observed in the past, the peak friction angle is reduced with the increase in sample size (Bandis 1990) according to the following equation:

$$\phi_p = \phi_b + d_n + S_A \tag{6.1}$$

where ϕ_p = peak friction angle, ϕ_b = basic friction angle, d_n = dilation angle, and S_A = scale-dependency of asperity failure component.

As the authors' model for unfilled joint does not include the effect of specimen size on the shear strength of joints, it is suggested that this effect should be incorporated under CNS testing in the future, and the model be revised accordingly.

However, for infilled joints the authors' model could still be used without significant modification for the prediction of joint strength.

6.2.6 *Extension in numerical modelling*

The overall deformation patterns and potential failure modes associated with an underground excavation or a slope in a jointed media can only be analysed with appropriate numerical techniques such as the Finite Element Method (FEM) or Discrete Element Method (DEM). Conventional Mohr-Coulomb and empirical Barton-Bandis parameters have often revealed significant inaccuracies in comparison with field observations, especially where water flow through joints takes place, hence effective stresses need to considered.

The authors' recommendation is to develop a series of computer subroutines to solve the governing equations representing the true shear resistance of field joints under CNS, including the role of irregular roughness, infill effects, scale effects, and internal water pressures. These subroutines could then be linked with UDEC or FLAC, for numerical solution.

REFERENCES

Bandis, S.C. 1990. Scale effects in the strength and deformability of rocks and rock joints. *Proc. Int. Workshop on Scale Effects in Rock Masses, Loen, Norway*, Balkema Publishers, Rotterdam, pp. 59-76.

Barton, N. 1987. Predicting the behaviour of underground openings in rock. *4th Manuel Rocha Memorial Lecture*, Lisbon.

Barton, N. & Choubey, V. 1977. The shear strength of rock joints in theory and practice. *Rock Mechanics*, 10(½): 1-54.

Herget, G. 1988. *Stresses in rock*. Balkema Publishers, Rotterdam, 179 p.

Hoek, E. & Brown, E.T. 1980. *Underground excavations in rock. Institution of Mining and Metallurgy*, London.

Infanti, N. & Kanji, M.A. 1978. In-situ shear strength, normal and shear stiffness determinations at Agua Vermelha Project. *Proc. Int. Congress of IAEG*, Buenos Aires, pp. 371-373.

Leichnitz, W. & Natau, O. 1979. The influence of peak shear strength determination of the analytical rock slope stability. *Proc. 4th Int. Congr. Of ISRM, Montreaux*, Vol. 2: 335-341.

Londe, P. 1973. The role of rock mechanics in the reconnaissance of rock foundations. *Qly J. Engng. Geol.*, Vol. 6(1).

Salas, I.A.J. 1968. Mechanical resistances. *Proc. Int. Symp. on Rock Mechanics, Madrid*, Theme II, pp.115-130.

Yoshinaka, R. & Yamabe, T. 1981. A strength criterion of rocks and rock masses. *Int. Symp. Weak Rock, Tokyo*, Vol. 2.

APPENDIX

Program code for shear strength model

Program code to predict unfilled and infilled joint shear behaviour using a model based on the Fourier series, energy balance principle and hyperbolic relationship.

```
'y = a0/2 + Σaₙ cos (2πnx/T) + Σbₙ sin (2πnx/T)        n=1-inf
'y = dilation, x = Shear displacement, T = period of cycle

CLS
CONST Pi = 3.141592
INPUT "Input Data File Name with path and extension"; FileDat$

DIM nlim AS INTEGER 'nlim= limit of n
INPUT "provide n limit"; nlim

DIM a(0 TO nlim) AS SINGLE, b(1 TO nlim) AS SINGLE

OPEN FileDat$ FOR INPUT AS #1

LINE INPUT #1, heading1$
LINE INPUT #1, heading2$
'Tstart = intial shear displacement, Tend = final shear displacement
'Tintvl = interval of measurement of dilation
'Splength = specimen length, Spwidth = specimen width, mm
'InNormStss = Initial normal stress (σₙₒ), Mpa, FricAngl = basic friction angle
'InAngle = initial asperity angle
INPUT #1, TStart!, TEnd!, TIntvl!, Spwidth, Splength, Stiffness, InNormStss, FricAngl, InAngle
NumDiv% = (TEnd! - TStart) / TIntvl!
DIM Dilation(0 TO NumDiv%) AS SINGLE
DIM EstDil(0 TO NumDiv%) AS SINGLE
DIM EstDilCorr(0 TO NumDiv%) AS SINGLE
DIM NormStss(0 TO NumDiv%) AS SINGLE
DIM ShearStss(0 TO NumDiv%) AS SINGLE
DIM Slope(0 TO NumDiv%) AS SINGLE
DIM SlopeCorr(0 TO NumDiv%) AS SINGLE

DilRead:
FOR x = 0 TO NumDiv%
 INPUT #1, Dilation(x)
NEXT x
'TAStart = initial t/a ratio, TAEnd = final t/a ratio
```

155

```
'Alfa and Beta = hyperbolic constants, Rf = tension cut-off
INPUT #1, Infillcod%
IF Infillcod% = 0 THEN GOTO EndRead
INPUT #1, TAStart, TAEnd, NumStep%, Alfa, Beta, Rf

DIM PeakShearDrop(0 TO NumStep%) AS SINGLE
DIM InfillPeak(0 TO NumStep%) AS SINGLE

EndRead: CLOSE #1

FourierCoeff:
REDIM a(0 TO nlim) AS SINGLE, b(1 TO nlim) AS SINGLE
FOR n = 0 TO nlim
  FOR k = 0 TO NumDiv% - 1
    a(n) = a(n) + Dilation(k) * COS(2 * Pi * k * n / NumDiv%)
  NEXT k
  a(n) = a(n) * 2 / NumDiv%
NEXT n

FOR n = 1 TO nlim
  FOR k = 0 TO NumDiv% - 1
    b(n) = b(n) + Dilation(k) * SIN(2 * Pi * k * n / NumDiv%)
  NEXT k
  b(n) = b(n) * 2 / NumDiv%
NEXT n

'GOTO plot

normalstress:
'Check and input dilation correction coefficients
IF InNormStss < .56 THEN
  BEEP: BEEP: PRINT : PRINT "Initial Normal Stress is below 0.56MPa"; CHR$(10); "for this
setup you need dilation correction"; CHR$(13)
  INPUT "Input Dilation Correction Coefficient File Name with path and ext."; FileCorr$

  OPEN FileCorr$ FOR INPUT AS #1

  LINE INPUT #1, headcorr1$
  INPUT #1, nlimcorr
  DIM ac(0 TO nlimcorr) AS SINGLE, bc(1 TO nlimcorr) AS SINGLE
  INPUT #1, ac(0)
  FOR n = 1 TO nlimcorr
    INPUT #1, ac(n), bc(n)
  NEXT n
  CLOSE #1
ELSE
END IF
'dilation correction end
T = TEnd! - TStart!: k = 1
FOR x = (TStart! + TIntvl!) TO TEnd! STEP TIntvl!
```

```
'Dilation
  y = 0
  FOR n = 1 TO nlim
    y = y + a(n) * COS(2 * Pi * n * x / T) + b(n) * SIN(2 * Pi * n * x / T)
  NEXT n
  y = a(0) / 2 + y
EstDil(k) = y
k = k + 1
NEXT
EstDil(0) = 0
'
T = TEnd! - TStart!: k = 1
FOR x = (TStart! + TIntvl!) TO TEnd! STEP TIntvl!
'Dilation
  y = 0
  FOR n = 1 TO nlim
    y = y + a(n) * COS(2 * Pi * n * x / T) + b(n) * SIN(2 * Pi * n * x / T)
  NEXT n
  y = a(0) / 2 + y

'dilation correction
  SELECT CASE InNormStss
  CASE IS < .56
    Ycor = 0
    FOR n = 1 TO nlimcorr
      Ycor = Ycor + ac(n) * COS(2 * Pi * n * x / T) + bc(n) * SIN(2 * Pi * n * x / T)
    NEXT n
    Ycor = ac(0) / 2 + Ycor
  CASE ELSE
    Ycor = 0
  END SELECT
EstDilCorr(k) = y - Ycor
NormStss(k) = InNormStss + Stiffness * EstDilCorr(k) * 1000 / (Spwidth * Splength)
k = k + 1:
NEXT
EstDilCorr(0) = 0
NormStss(0) = InNormStss

ShearStress: CLS

ShearStss(0) = 0: PeakShear = 0: NormAtPeak = InNormStss
FOR k = 1 TO NumDiv%
  Slope(k) = ATN((EstDil(k) - EstDil(k - 1)) / TIntvl!)
'Shear model Eqn. 7.9b
ShearStss(k) = NormStss(k) * (TAN(FricAngl * Pi / 180) + TAN(InAngle * Pi / 180)) / (1 - TAN(Pi
* FricAngl / 180) * TAN(Slope(k)))
'Shear model Eqn. 7.7
'ShearStss(k) = NormStss(k) * TAN(FricAngl * Pi / 180 + Slope(k))
  IF PeakShear < ShearStss(k) THEN
      PeakShear = ShearStss(k)
```

```
      NormAtPeak = NormStss(k)
      DisAtPeak = k * TIntvl! + TStart!
  ELSE
  END IF
NEXT k


Infill:
IF Infillcod% = 0 THEN GOTO Printing
TAIntvl = (TAEnd - TAStart) / NumStep%
FOR k = 0 TO NumStep%
  PeakShearDrop(k) = InNormStss * (TAStart + TAIntvl * k) / (Alfa * (TAStart + TAIntvl * k) +
Beta)
  InfillPeak(k) = PeakShear - PeakShearDrop(k)
  MaxDrop = PeakShear - (InNormStss * Rf / Alfa)
  IF InfillPeak(k) < MaxDrop THEN InfillPeak(k) = MaxDrop
NEXT k


PRINT "Do you want to estimate Infill peak at any T/A? (Y/N) ": DO: Inest$ = INKEY$: LOOP
UNTIL Inest$ <> ""
IF UCASE$(Inest$) <> "Y" GOTO Printing
INPUT "Give T/A value "; TA
  PeakShearDrop = InNormStss * TA / (Alfa * TA + Beta)
  InfillPeak = PeakShear - PeakShearDrop


Printing:
PRINT "print result in file (Y/N) ? ": DO: prin$ = INKEY$: LOOP UNTIL prin$ <> ""
IF UCASE$(prin$) = "Y" THEN GOTO printfile


printscreen:
CLS


PRINT "Fourier Coefficients"


PRINT "        a        b"
PRINT
PRINT USING " ( 0)  ###.#####"; a(0)


FOR n = 1 TO nlim
  PRINT USING " (##)  "; n;
  PRINT USING " ###.#####"; a(n);
  PRINT USING " ###.#####"; b(n)
NEXT n


PRINT "press any key": DO: key$ = INKEY$: LOOP UNTIL key$ <> ""
CLS
PRINT
PRINT " Disp(mm)  Dil(mm)   NrSt(Mpa)  Slop(ø)  ShrSt(Mpa)"
FOR k = 0 TO NumDiv%
PRINT USING "####.##"; TStart! + k * TIntvl!;
PRINT USING "####.#####"; EstDil(k);
```

```
PRINT USING "  ####.#####"; NormStss(k);
PRINT USING "  ####.##"; Slope(k) * 180 / Pi;
PRINT USING "  ####.#####"; ShearStss(k)
IF k > 0 THEN IF (k / 20) - INT(k / 20) = 0 THEN PRINT "press any key": DO: key$ = INKEY$:
LOOP UNTIL key$ <> "": CLS
NEXT k
PRINT USING "Peak Shear ####.#### Mpa "; PeakShear
PRINT USING "Normal at Peak Shear ####.#### Mpa "; NormAtPeak
PRINT USING "Displacement at Peak Shear ###.## mm "; DisAtPeak

PRINT "press any key": DO: key$ = INKEY$: LOOP UNTIL key$ <> "": CLS

IF Infillcod% = 0 THEN GOTO ScrPrintEnd
PRINT
PRINT "  T/A    PeakShearDrop    InfillPeak "
 FOR k = 0 TO NumStep%
 PRINT USING "###.## "; (TAStart + TAIntvl * k);
 PRINT USING "  ####.#####  "; PeakShearDrop(k);
 PRINT USING "  ####.#####  "; InfillPeak(k)
 IF k > 0 THEN IF (k / 20) - INT(k / 20) = 0 THEN PRINT "press any key": DO: key$ = INKEY$:
LOOP UNTIL key$ <> "": CLS
 NEXT k

PRINT "press any key": DO: key$ = INKEY$: LOOP UNTIL key$ <> "": CLS

IF UCASE$(Inest$) <> "Y" GOTO ScrPrintEnd
 PRINT USING " PeakShearDrop = ####.#### at T/A = ##.## "; PeakShearDrop; TA
 PRINT USING " InfillPeak = ####.#### at T/A = ##.## "; InfillPeak; TA

ScrPrintEnd:
GOTO printend

printfile:
INPUT "Type result file name"; Resfile$

OPEN Resfile$ FOR OUTPUT AS #2
printnow:
PRINT #2, "Fourier Coefficients for:"
PRINT #2, heading2$: PRINT #2,
PRINT #2, "      a        b"
PRINT #2,
PRINT #2, USING "  ( 0)  ###.#####"; a(0)

FOR n = 1 TO nlim
 PRINT #2, USING "  (##) "; n;
 PRINT #2, USING " ###.#####"; a(n);
 PRINT #2, USING " ###.#####"; b(n)
NEXT n
PRINT #2,
PRINT #2, " Disp(mm)  Dil(mm)   NrSt(Mpa)  Slop(ø)  ShrSt(Mpa)"
```

```
FOR k = 0 TO NumDiv%
PRINT #2, USING "####.###"; TStart! + k * TIntvl!;
PRINT #2, USING "####.#####"; EstDil(k);
PRINT #2, USING " ####.#####"; NormStss(k);
PRINT #2, USING "  ####.##"; Slope(k) * 180 / Pi;
PRINT #2, USING "  ####.#####"; ShearStss(k)
NEXT k
PRINT #2, USING "Peak Shear ####.#### Mpa "; PeakShear
PRINT #2, USING "Normal at Peak Shear ####.#### Mpa "; NormAtPeak
PRINT #2, USING "Displacement at Peak Shear ###.## mm "; DisAtPeak

IF Infillcod% = 0 THEN GOTO FilePrintEnd
PRINT #2,
PRINT #2, " T/A    PeakShearDrop    InfillPeak "
 FOR k = 0 TO NumStep%
 PRINT #2, USING "###.## "; (TAStart + TAIntvl * k);
 PRINT #2, USING "  ####.##### "; PeakShearDrop(k);
 PRINT #2, USING "   ####.##### "; InfillPeak(k)
 NEXT k

IF UCASE$(Inest$) <> "Y" GOTO FilePrintEnd
 PRINT #2, USING " PeakShearDrop = ####.#### at T/A = ##.## "; PeakShearDrop; TA
 PRINT #2, USING " InfillPeak = ####.#### at T/A = ##.## "; InfillPeak; TA

FilePrintEnd: PRINT "Thank you- Look at your result file: "; Resfile$; "  Calculated from data file:
"; FileDat$
CLOSE #2

printend:
END
```

TYPICAL INPUT DATA FILE

```
Dilation and Shear Displacement data for
Joint type: 1   Normal stress: 0.56 MPa
0 30 .5
75 250
8.5 .56 37.5 9.5
0 .05 .13 .19 .25 .3 .35 .41 .45 .5 .55 .6 .65 .69 .74 .78 .83 .87 .91 .95 .98 1.01 1.04 1.07 1.09 1.11
1.13 1.15 1.16 1.17 1.18
1.17 1.16 1.15 1.13 1.11 1.09 1.07 1.04 1.01 .98 .95 .91 .87 .83 .78 .74 .69 .65 .6 .55 .5 .45 .41 .35 .3
.25 .19 .13 .05 0
1 0 2 10 .71 .18 .85
```

DATA format:
start, end, interval of series
specimen width (mm) and length (mm)
stiffness (kN/mm), Initial Normal Stress (MPa), Friction Angle (deg),
and Initial asperity angle (deg).

All Dilations
Infillcode(0 = no. Other = yes), T/A start, T/A End, Number of Step, α, β, reduction factor (R_f).

TYPICAL OUTPUT FILE

Fourier coefficients for:
Joint type: I   Normal stress: 0.56 MPa

	a	b
(0)	1.44667	
(1)	−0.49320	0.00000
(2)	−0.07689	0.00000
(3)	−0.04259	0.00000

Disp (mm)	Dil (mm)	NrSt (Mpa)	Slop (ø)	ShrSt (Mpa)
0.000	0.00000	0.56000	0.00	0.00000
0.500	0.11711	0.61309	3.18	0.69859
1.000	0.13621	0.62175	2.19	0.59867
1.500	0.16703	0.63572	3.53	0.62369
2.000	0.20816	0.65436	4.70	0.65282
2.500	0.25776	0.67685	5.67	0.68476
3.000	0.31373	0.70222	6.39	0.71801
3.500	0.37381	0.72946	6.85	0.75106
4.000	0.43581	0.75757	7.07	0.78254
4.500	0.49771	0.78563	7.06	0.81137
5.000	0.55777	0.81286	6.85	0.83690
5.500	0.61469	0.83866	6.49	0.85889
6.000	0.66759	0.86264	6.04	0.87753
6.500	0.71607	0.88462	5.54	0.89329
7.000	0.76016	0.90460	5.04	0.90686
7.500	0.80023	0.92277	4.58	0.91900
8.000	0.83694	0.93941	4.20	0.93046
8.500	0.87109	0.95489	3.91	0.94187
9.000	0.90349	0.96958	3.71	0.95367
9.500	0.93488	0.98381	3.59	0.96608
10.000	0.96579	0.99782	3.54	0.97908
10.500	0.99648	1.01174	3.51	0.99239
11.000	1.02693	1.02554	3.48	1.00552
11.500	1.05678	1.03907	3.42	1.01782
12.000	1.08542	1.05206	3.28	1.02853
12.500	1.11201	1.06411	3.04	1.03691
13.000	1.13560	1.07481	2.70	1.04233
13.500	1.15522	1.08370	2.25	1.04434
14.000	1.16997	1.09039	1.69	1.04275
14.500	1.17913	1.09454	1.05	1.03761
15.000	1.18223	1.09594	0.36	1.02925

Peak shear stress   1.0443 MPa
Normal stress at peak shear 1.0837 MPa
Displacement at peak shear   13.50 mm

T/A	Peak shear drop	InfillPeak
0.00	0.00000	1.04434
0.20	0.34783	0.69652
0.40	0.48276	0.56158
0.60	0.55446	0.48989
0.80	0.59893	0.44541
1.00	0.62921	0.41513
1.20	0.65116	0.39318
1.40	0.66780	0.37654
1.60	0.68085	0.37392
1.80	0.69136	0.37392
2.00	0.70000	0.37392

Subject index

Bolted joints 20, 144-147, 153

Coefficients
empirical 48, 90
Fourier 103-104, 110-111
hyperbolic 114-116
joint roughness 2-5, 100-101
Constant Normal Stiffness (definitions) 18-20

Deformation (see Dilation)
effect of boundary condition 31, 50, 63-64, 141
effect of shear rate 28, 49-50
effect of stiffness 28, 131-132
elasto-plastic deformation 106
plastic (yield) deformation 131
shear displacement rate 29, 31
Dilation
angle 47, 99-100, 107, 132
conceptual model 99-101, 103
dilation/asperity height ratio 48
effect of horizontal displacement 40-41, 100
effect of normal stress 27, 30, 39-40, 48
effect of roughness 100-101
Fourier analysis 15, 109

infilled joint dilation 75-78
initial normal stress 48
model predictions 120-123
peak dilation angle 45
saw-tooth (triangular) joints 29, 34-35
Discontinuities (joints and fractures)
aperture closure and opening 99
asperities 45-46, 51, 79, 97
asperity angle 45-47, 58, 105-106, 111
basic friction angle 25, 44-45, 96, 105, 111, 131, 153
behaviour of clean rock joints 25-28, 36-37
behaviour of field joints (natural) 24-26, 37-38, 40-42
cohesion 106
dilation 27, 30, 34-35, 39-41, 48, 76, 99-101, 103-105, 120-123
effect of initial normal stress 33-34, 36-41, 47, 68-69, 76-78, 85
fracture surface measurement 11- 14
friction angle 44-47, 66, 95-97, 100-102, 105-107, 131
joint interfaces 27
joint roughness coefficient (JRC), 2-5,

45-46, 100-101
joint wall compressive strength (JCS) 100
model materials 23
natural field joints 24-26, 40-42
properties of jointed rocks 15-16, 23
relative thickness 57-62
residual friction angle 99-101
rock joint testing 25-45
saw-tooth joints 23, 32, 103
soft rock joints 29-37
surface profiles 14
unfilled joints 36-45
uniaxial compression strength 25, 46

Excavations 18, 140-143

Factor of safety 147-148
Fractures (see Discontinuities)

Infilled joints
bentonite 70
boundary condition 50, 63-64
consolidation 68
drainage 63
factors controlling strength 55-56
influence of infill 54-55
interaction with rock 64-65
kaolin 58-59
material selection 70

163

over-consolidation ratio
(OCR) 90
preparation 72-73
shear behaviour 74-91
shear strength modelling
87-90, 113-115
stress ratio 89
testing 68-74
thickness 57-62, 66,
77-85

Joints (see Discontinuities)

Laboratory testing
displacement measure-
ment 22
infilled joint testing 68-74
natural field joints 24-26,
37-38, 40-42
rock joint testing 25-45
shear apparatus 21-22
shearing rate 28-31, 50
stress paths 42-43, 81-83
surface profiling equip-
ment 11-13
Loading concepts and applications
bolted rock joint 20
constant normal load
(CNL) 18-20, 147-148
constant normal stiffness
(CNS) 18-20, 26, 142,
146
ground reaction curve
142-143
practical applications
19-20, 143-147
rock socketed piles 19

Model simulation (see Nu-
merical analysis)
energy balance 94-96
Fourier series 102-103,
107-109
fractals 4-8
graphical 98
hyperbolic 113- 117
mechanistically based
models 97-100
model requirements 102

shear strength models
87-90, 102-117

Numerical analysis (see
Model simulation)
block discretisation 133
continuous yielding
131-132
distinct element model-
ling 127- 129,
134- 137, 140-142
predictions 119-126
UDEC analysis 130,
133-137

Rock
bolt applications 20,
144-145
slope analysis 144-147
Roughness
characterisation 2-15
Fourier transform 14-15
fractal parameters 6-8
fracture surface profile
12, 14
joint roughness coeffi-
cient 2-5, 100-101
measurement 2, 11- 15
parameters 97
random profile generation
5-8
self-affine profile 5
semi-variogram functions
8, 10
spectral method 9-11

Scale effects 153
Shear strength (peak shear
stress)
Barton's model 45-46,
112-113
Barton-Bandis 130
bi-linear 105
CNS envelopes 86
effect of boundary condi-
tion 50-51
effect of infill 83-86,
113-115
effect of shearing rate 50
empirical models 45-48

Fourier model 102-105,
107-109
influence of infill 54-55,
83
Johnston and Lam 94-95
Ladanyi and Archambault
model 93
Mohr-Coulomb 70, 130
natural joints 37-38
normalised strength drop
(NSD) 83, 114-115
Patton's model 45, 50,
105-106
strength envelopes 35, 42,
44-45, 50, 70, 86, 126-
127
Slope stability 144-147
Statistical (probability)
methods 3, 6, 9-12
Stiffness
bolt-grout 144
conceptual model 18, 132
external stiffness 67-68
normal stiffness 18, 26,
28-29, 100, 131
shear stiffness 131
Stress conditions
boundary condition
31-33, 50-51
effect of horizontal dis-
placement 31-33
effective stresses 152-153
initial normal stress
33-34, 36-41, 47,
68-69, 76-78, 85, 127,
130
lateral confinement 68
normal stress response
27, 30-31, 33-34, 68,
119-123
peak shear stress 83-87,
99
shear stress response
29-30, 32-33, 124-129
stress paths for clean
joints 42-43
stress paths for infilled
joints 81-83